Telecommuting and Virtual Offices: Issues and Opportunities

Nancy J. Johnson
Capella University, USA

IDEA GROUP PUBLISHING
Hershey USA • London UK

Acquisition Editor: Mehdi Khosrowpour
Managing Editor: Jan Travers
Development Editor: Michele Rossi
Copy Editor: Nicholas Tonelli
Typesetter: Tamara Gillis
Cover Design: Deb Andre
Printed at: Sheridan Books

Published in the United States of America by
        Idea Group Publishing
        1331 E. Chocolate Avenue
        Hershey PA 17033-1117
        Tel: 717-533-8845
        Fax: 717-533-8661
        E-mail: jtravers@idea-group.com
        Web site: http://www.idea-group.com

and in the United Kingdom by
        Idea Group Publishing
        3 Henrietta Street
        Covent Garden
        London WC2E 8LU
        Tel: 171-240 0856
        Fax: 171-379 0609
        Web site: http://www.eurospan.co.uk

Library of Congress Cataloging -in-Publication Data

Telecommuting and virtual offices : issues and opportunities / [edited by] Nancy J.
Johnson.
        p.com.
    Includes bibliographical references and index.
    ISBN 1-878289-79-9 (paper)
        1. Telecommuting. 2. Virtual reality in management. I. Johnson, Nancy J., 1951-

HD2336.3 .T44 2000
658.3'128--dc21   00-047192

British Cataloguing in Publication Data
A Cataloguing in Publication record for this book is available from the British Library.

Telecommuting and Virtual Offices: Issues and Opportunities

Table of Contents

Section Three: EMPLOYEE ISSUES

Preface

In assembling a book about telecommuting, the scope of the stakeholders offers many venues for analysis and exploration. The growth of employees interested in telework is not limited to the US, and is growing as quickly as the technology advances in portable computing and telecommunications. Communities, employers and employees are all venturing into uncharted territories of working independently and away from corporate center face-to-face interaction and support. Learning curves are steep, grounded in experiences that are examined and refined, as well as learning from other organizations that have tried it and are still doing it. Employees are now routinely asking for the option as a condition of employment.

The definition of telecommuting is elusive at best. Each organization and employee has a different perception of the characteristics of the structure of it as well as a myriad of names for it: telematics, virtual offices, mobile workers, home workers, and more. Does it mean working at home a few days a week or completely? Does it mean working out of a hotel or car while visiting customer sites? How is the nature of work output changed for an information worker as compared to the industrial era model of the employee chained to the means of production (e.g., a loom or a lathe)? Many managers are still grounded in managing by walking around (MBWA) and are extremely uncomfortable with the notion that a worker could be productive working out of view. Yet, this discomfort exists for managing information workers in the office too and managers are slow to shift evaluation paradigms and management styles.

Employers are faced with unexplored issues of responsibility and liability for offsite employees' health, as well as deploying thousands of dollars of equipment to employees traveling through airports and in vehicles. Servicing the equipment, protecting the data on it and ensuring ease of use/quick problem resolution for employees pushes against the assumptions of ease associated with stewardship of in-office employees and equipment. Ergonomic issues arise in any work setting, and access security to databases is a threat all the time. Labor unions are seeing potential for exploitation of employees, as well as inconsistent organizational expectations for teleworkers.

Communities are challenged to reduce traffic/parking congestion, improve the quality of life in residential areas, reduce pollution from cars, and attract competent workforces for local employers. Telecommuting offers reductions to many of these concerns, but requires communities to examine/upgrade zoning ordinances regarding working out of homes, traffic patterns, public transport options, telecommunication infrastructures (or lack of them), and support networks in the communities for teleworkers.

The authors in this book provide answers and guidance for many of the problems associated with introduction of telecommuting, and suggest many topics for more research. Learning from experience has been the mode for many employers and employees, and examination of these experiences with the rigor of academic study benefits all concerned. An academic perspective in the investigation is appropriate because professors have been 'accidental telecommuters' for many years (in the sense that Anne Tyler, in her book *The Accidental Tourist,* used to describe the traveler who didn't want to travel in the first place but had to for work). Academics quickly learn that a home office is necessary to have a peaceful setting for writing, especially at non-standard working hours and frequently have better computer equipment than the university, due to budget constraints, can provide. Academics frequently work in collaboration with colleagues far distant, and were early adopters of the Internet to share information and papers. Academics are also being driven to offering classes in a distance education delivery model to meet the needs of nontraditional students who are working from their own homes or from the road, at non-standard class hours.

The practitioner perspective in research is also important because the world of theory only has tangential connections to the reality of today's workplace: working smarter with fewer employees, globalization of multinational firms, serving the customer's needs (not the employee's wishes), keeping up with the accelerating pace of technological change and finding and keeping good employees in a tight labor market. The employer's voice of experience is critical in balancing the academic studies of different aspects of telecommuting.

This book presents the best work of a wide variety of authors and styles: practitioners and academics, US and internationally based, formal studies and case studies and information from the perspective of the community, the employer and the employee.

In the community section, the legal aspects of offering the programs are explored by Baruch and Smith. While based in England, the issues are universal. The potential for telecommuting in Ireland contrasted to the actual rate of adoption is discussed by Adam and Crossan. Establishing and running a national promotion center for telework in the Netherlands is presented by de Bruin. Tackling the thorny issue of using public telephone network infrastructure for telecommuters is done by Bumblis.

Henquinet provides guidance on the selection of the right employee for telework, and Platt and Page provide tips for managing the virtual team of telecommuting employees. Harrington and Ruppel examine organizational style as a factor in measuring the potential for successful telecommuting programs. St. Amant addresses the awareness needed for effective intercultural communication in global organizations. Johnson provides a case study of an international insurance firm that has refined its virtual office program over six years of use.

From the perspective of the employee, Pinsonneault and Boisvert review the effects of telecommuting on both the individual and the organization. Staples addresses improving the effectiveness of offsite workers, and Rodstein and Watters describe ergonomic issues for all employees. McCloskey skewers the myths that have evolved about telework through a formal study.

The authors open a Pandora's box of many more issues ripe for further study and examination. Through sharing of knowledge in formal and informal channels, employers, employees and communities will increase their effectiveness in supporting this new paradigm of working environments.

Special thanks to Patricia L. Gregory for her work with the authors in editing each chapter in the early drafts. Her assistance was invaluable.

Nancy Johnson
Editor

COMMUNITY ISSUES

Chapter I

Telecommuting and the Legal Aspects

Ian T. Smith and Yehuda Baruch
University of East Anglia, UK

INTRODUCTION

This chapter discusses teleworking/telecommuting from a legal perspective, as applied to the management of teleworkers. The main issues covered are the definition of teleworking, employment relationships and employment contracts for telecommuting, health and safety implications of teleworking, and other legal-related considerations to be taken. Lastly, implications are discussed for both the management of organizations and the legal establishment.

This chapter examines teleworking from a legal point of view. Although it was written from a European (in particular, British) perspective, we have tried to use material with relevance extending beyond the boundaries of one specific country, and the arguments are intended and expected to apply to a much wider audience and circumstances.

THE EMERGING IMPORTANCE OF TELEWORKING

Already in the 1950s the literature on technological change suggested that telecommunications, combined with computing technology, could enable work to be relocated away from the traditional office (Jones, 1957-58). Widespread interest in teleworking started in the 1970s, when the term "teleworking" first came into usage to indicate remote working from the office (Nilles, Carlson, Gray & Hanneman, 1976). Today, interest in teleworking is still growing among employees, employers, policy-makers (e.g. transportation planners), communities, the telecommunications industry and many others (Handy and Mokhtarian, 1996). The effective use of teleworking serves as a base for the 'virtual organization' (Davidow & Malone, 1992; Chesbrough & Teece, 1996; Peiperl & Baruch, 1997). As happens whenever a novel type of work arrangement is introduced, teleworking challenges the legal system with new issues and questions. More people and organizations are facing situations and problems that have no precedents from which to learn.

The aim of this chapter is to provide guidelines and advice to employers and employees, particularly in relation to two main constructs: employment contracts, and health and safety aspects of teleworking.

WHAT IS TELEWORKING? PROBLEMS IN DEFINITION AND MEASUREMENT

Although teleworking has been discussed for many years, a universal definition still is not in place (IRS, 1996; Moon & Stanworth, 1997). There is not even an agreed term: "Teleworking," "telecommuting," "working-at-home," (or "home-working"), "working-at-a-distance," "remote work," and recently "virtual work" are among the terms that have been used to cover different working policies and practices. A variety of definitions for teleworking exist, causing problems of ambiguity, which increase the difficulties for the legal system to deal with the phenomenon. Lack of agreed definition prevents a coherent approach in a way that, for example, precedents based on one type of teleworking may not be relevant for a different type of teleworking. Another part of the problem refers to the title used, but there is agreement on the issue of the technology involved. Teleworking uses electronic media as its main 'tool' (Mitchell, 1995; Negroponte, 1995).

Teleworking is usually defined, first, in terms of location and, second, in terms of technology in use. In a pioneering work in this area, Shamir and Salomon (1985) defined teleworking as "working at home." However, teleworking is not limited to a single location, and certain types of work (e.g. the 'Rug-industry' – home-based sweatshops) are not perceived as teleworking. Grant (1985) described teleworking as "a kind of remote working, or doing normal work activities while away from one's normal workplace," which covers a wider range but still falls short of stating what constitute "one's normal workplace." Others added later the technology aspect (Cross & Raizman, 1986; Olson, 1988) or time frame e.g. "some or all of the time" (Huws, Korte & Robinson, 1990).

The European Commission (1994) and Korte and Wynne (1996) argued that teleworking comprises at least three main elements: (a) location of the workplace, which means it is partially or fully independent from the location of the employer, contractor, client, etc.; (b) use of information technology (IT), mainly personal computers, e-mail, faxes and telephones; and (c) organizational form and communication link to the organization. In an approach more relevant to the legal dimension of teleworking, Trodd (1994) acknowledged that there are several forms of teleworking: individual teleworking, corporate teleworking, executive teleworking and contract teleworking. Teleworkers may operate in various locations: at home, a satellite office, telecottage or mobile teleworking.

Examining a number of different definitions of teleworking, we will use a simple and general definition for practical and academic use (still readers should be aware that differences in definitions may have implications for legal aspects concerned with teleworking): Teleworking is a mode of work in which employees perform all or a significant part of their roles from a base physically separated from the location of their employer (usually their home), and use information technology as their main tool for operation and communication.

Again, we do not refer to self-employed people, as their legal status is different from that of employees, and much has been written on the legal status of self-employed.

How Many Telework?

Identifying the number of teleworkers can vary by the threshold of time spent at home (compared to the time spent at the employer's premises) which distinguish between teleworkers and non-teleworkers: Does teleworking mean only cases where people work all their time at home? More than half the time? At least one or two days a week? From a legal viewpoint, there is a difference between occasional or part-time teleworking and full-time teleworking. This difference can be tested by availability of office space at home and at the employer's premises. Another question is whether self-employed people should be included in the teleworking population. A distinction also can be made between people working from home, telecottages, and mobile workers. Qvortrup (1998) echoed Huws, Korte, and Robinson (1990) who warned against comparisons conducted using different base definitions. His example shows how, using different surveys within the same country (UK), the numbers varied considerably from 110,000 to 1,224,000 in 1992-1994 (Qvortrup 1998, p. 29). Only two years later, the IDS 616 (1996) claimed the actual number for the same population was between half a million and 1.7 million, *depending on the precise definition used.*

One difference that may be less relevant to the management of teleworkers, but is extremely important for legal reasons, is whether the teleworking takes place as an informal, voluntary occurrence, or whether it is a formal organizational practice.

What is undisputed is that the teleworking phenomenon is widespread, is increasing steadily, and is expected to continue. Teleworking fits well with the new set of career aspirations that characterize the present generation, in particular for females entering the labor market and as more males recognize the importance of a balanced life (cf. Arthur, Inkson & Pringle, 1999).

Teleworking provides a unique solution for both employers and employees, but it is a complex and complicated mode of work characterized by ambiguous boundaries between work, family, employment and leisure. Traditional control mechanisms do not necessarily apply for teleworking, creating a situation that calls for a different interpretation by the legal system.

TELEWORKING AND THE LAW

While much attention has been paid to work arrangements, work attitudes toward and performance implications of teleworking, the literature lacks a discussion of the legal aspects and implications of teleworking. With the contemporary trend toward a 'litigious society' many managerial concerns focus on the legal considerations of processes and practices. Thus it is surprising how little attention has been paid to the legal aspect in teleworking literature. This chapter will focus on two significant legal elements of teleworking: employment contracts and the health and safety aspects of teleworking.

EMPLOYMENT CONTRACTS AND TELEWORKING

The employment relationship is explained in common law jurisdictions in terms of the employment contract, the extent and terms of which are used by courts and tribunals to govern that relationship legally. Not surprisingly, the basic law on contracts of employment grew (in the late nineteenth century[1] and the first three-quarters of the twentieth century) in the context of manual full-time male employment in agriculture or manufacturing, with a clear hierarchical structure of management and control, and at a time when the demarcations between master and servant, and between the self-employed professional and the hired worker, were far more obvious. The last quarter of the twentieth century has, however, seen enormous changes in industrial organization and flexibility such as the rise of service and information technology industries and of female employment, to the extent that the old model for the employment contract is no longer dominant. What used to be rather dismissively referred to as 'atypical employment' has in many areas become the norm, especially in those industries that have seen the bulk of their job creation in the late 1980s and 1990s. The literature on this is, of course, extensive. For official sources in the UK, see Wareing (1992) *Employment Gazette*, 55, 88, 225; Beatson (1995); and the previous UK Conservative government's White Paper titled *People, Jobs and Opportunity* (Cm 1810, 1992). Also Nolan and Walsh (1995) referred to the structure of the economy in relation to labor in Britain. It has, therefore, been essential for the law to keep pace with these developments to ensure that these new and evolving forms of employment are covered by employment rights, both statutory and contractual.

At the forefront of these important developments has been that ancestor of the modern teleworker, the 'homeworker' or 'outworker.' Although in some ways very different (in particular often involving repetitive manual work, such as finishing clothing [Stanworth & Stanworth, 1989]), outworking has set at least some of the agenda, with the UK courts holding that outworkers could in law be 'employees' and so entitled to employment rights. This was the case when a homeworker who made heels for shoes was held to be an employee and so able to claim unfair dismissal (*Airfix Footwear Ltd. v Cope*, 1978), and when the homeworker who sewed pockets into trousers was held to be an employee, a case approved by the Court of Appeal in *Nethermere (St. Neots) Ltd. v Gardiner* (1984). The application of such existing developments to more modern forms of teleworking raises the following legal points.

Determination of Employment Status

Originally the test for whether an individual was an employee or self-employed was the 'control test,' i.e., whether employers could control not just what individuals did, but also the way they did it. This naturally presupposed a high level of supervision, usually on the employer's own premises, and often allied to a rigid system of hours, breaks and payments. The position of the teleworker could be said to be the antithesis of the control test, and so on that basis would start off as being prima facie a case of self-employment. However, the control test was an early casualty of diversified employment patterns after World War II, especially the increasing number of professional people in employment relationships (e.g., doctors, surgeons, state lawyers) where it could not realistically be said that the modus operandi was directly under the employer's control (Smith & Wood, 1996).

Teleworking requires a high level of trust and a supportive organizational culture to flourish (Baruch & Nicholson, 1997). However, contemporary work design and career systems in organizations are considered to be boundaryless, and the so-called alternative work arrangements use nontraditional modes of work, where teleworking is one of the notable ones (Arthur & Rousseau, 1996). In contrast to the fast-changing nature of work arrangements, relevant case law and legislation develops to a much slower pace.

The current common law approach to defining employment tends to be the 'pragmatic' or 'multiple' test, namely weighing *all* relevant factors in the relationship (form of hiring, method of payment, elements of control, organization of the work, provision of tools and equipment, power to delegate, method of termination) and determining where the balance of those factors lies. Although this is a much broader test, it still could raise strong arguments that a particular teleworker was self-employed. Moreover, there is a second hurdle for teleworkers who seek legal protection—in any particular state or jurisdiction there may be requirements of qualifying *continuous* service by the individuals before they can claim a legal right (even once it is clear that they are 'employees,' in the UK an employee must work for the employer continuously for two years before claiming a statutory redundancy payment and, since it was lowered in June 1999, for one year before being able to claim unfair dismissal). There is a danger that, if the work done is in some way sporadic, the employer might argue that the individual was an employee, but on separate noncontinuous employment contracts each time.

A good example in the UK was *Hellyer Bros Ltd. v McLeod*, (1987), where the Court of Appeal held that trawlermen working over a period of eight years for the same fishing company were in fact engaged on separate contracts each time they went on a fishing voyage, so when the work finally ceased, they had no statutory redundancy rights. To counter both of these hurdles, UK courts have developed broader concepts capable of bringing the 'atypical' worker into employment. In particular they have included an important factor called 'mutuality of obligations' and the idea of a 'global' or 'umbrella' contract. The former can mean that a relatively loose arrangement between employer and individual can, after a period of time, assume the sort of de facto mutuality normally expected in an employment relationship; for example, showing that the employer impliedly undertook to supply the individual with a reasonable amount of the work available and the individual impliedly undertook to do a reasonable amount of the work on offer.

The latter (the idea of a 'global' or 'umbrella' contract) can be used to link a series of apparently separate contracts into one overall employment contract, thus providing the necessary continuity of employment. In addition, a particular state or jurisdiction's employment legislation may have statutory rules deeming there to be continuity in some or all cases. Both of these developments could be used by a teleworker where the arrangement with the employer is not just to work at home, but also for the individual to have more control or discretion over the amount and timing of the work to be done.

One aspect of the future controversies over teleworkers' status that has become increasingly clear over recent years is that, in the UK at least, an employer cannot expect simply to apply the magic phrase 'casual worker' to a teleworker and thereby avoid all employment rights. The common law tests previously mentioned have been used to attack this easy assumption, though it has to be accepted that these developments cannot always be relied on, and cannot be taken too far. Thus, in *Carmichael v National Power* (2000) tourist guides employed on a 'casual, as required' basis (over a long period of time, but always on a relatively informal basis) were held by the House of Lords *not* to be employees, and so unable to claim employment protection rights. In light of this important case, it is therefore significant that the current Labour government's recent legislation has tended to be applied on a wider basis anyway: instead of applying only to 'employees,' it applies to anyone under a contract of employment *or* under any other contract *personally* to execute work or services for another person (other than in a professional capacity)[2].

Again, this could be useful for teleworkers – provided the personal service element was present, they could claim rights under this legislation even if doubt remained whether they were employees under the classic tests. Moreover, within the European Union, atypical workers are increasingly being protected by directives issued under the 'social Europe' provisions of the treaties, which of course have to be implemented in all the member states. At the time of this writing the *Part-Time Workers Directive (97/81/EC)* is about to come into force, banning discrimination against part-time workers, and a further directive giving similar protection to temporary workers is awaiting implementation. A teleworker whose contract was on a part-time or temporary basis could claim this protection just as much as a more traditional worker.

Modern Statutory Intervention

The UK Conservative government of 1979-1997 sought to follow the American model of a flexible, deregulated labour market, leaving employment terms as primarily matters of contract and market forces. While the current Labour government has espoused large parts of this approach, there now are some trends toward reregulation by laws that will inherently prove more difficult to apply to teleworkers. As a matter of domestic policy of the present government, the UK now has for the first time a national minimum wage. In the UK, separate minimum wages used to be laid down for certain non-unionized trades by UK Wages Councils, whose history goes back to 1909, as attempts to prevent exploitation of workers in the so-called 'sweated trades.' Significantly, the legislation expressly extended protection to contingent workers. However, this legislation and its councils were abolished in 1993 as a deregulatory measure. Although the new national minimum wage is set at a low level (the UK normal rate being £3.60 per hour) and largely aimed at low-paid manual labourers (including work traditionally done by outworkers), there are problems applying legislation such as this to teleworkers. They may seem to be remunerated at well above the minimum rate, *but* in order to determine this, teleworkers' actual hourly rate must be calculated. If they are paid on an hourly rate for work done, there is little problem, but as soon as there is payment on a salary or piecework (i.e., work-done) basis, the problem arises of how to determine their actual hours (by which to divide their income) in order to work out their hourly rate. This can be difficult enough when the individual works on the employer's premises, but if one aim of teleworking is to provide employees with more flexible work patterns, this could be a particular difficulty. One way that UK legislation seeks to resolve these difficulties is by allowing employer and employee to agree in advance on a realistic average number of worked hours, this number to be used in any minimum wage calculation.

The other source of reregulation is, of course, the EU. The principal manifestation of this so far has been the *Working Time Directive (93/104/EEC)*, enacted in the UK as the *Working Time Regulations 1998*. These seek to establish a normal weekly maximum of 48 hours per week, restrictions on night working, mandatory rest periods and a minimum of four weeks paid holiday per year. They are not restricted to full-time workers or to those working on the employer's premises or directly under the employer's control. At present, there are many loopholes and exceptions (negotiated into the Directive as permissible 'derogations' by the previous Conservative government) so that, for example, an individual can opt out in writing of the 48-hour maximum, and a trade union or workforce agreement can be used to amend or even disapply the night working and rest break provisions. However, the onus remains on the employer to show that, one way or another, the Regulations are being observed. Once again, a regime such as this would be much easier to organise when the employee is working on the employer's premises. In the case of teleworkers, compliance may require record keeping of hours worked (especially where the teleworker is paid on a salary basis, rather than on an hourly rate) that is, in fact, contrary to the increased flexibility and informality that the teleworking may well have been designed to encourage. This may be bad enough under the present regime, but at the time of this writing there are moves within the EU toward another Directive removing some or all of the derogations, in which case this highly regulatory structure would really start to bite, placing even more stringent requirements on employers of teleworkers within the EU.

One further area of EU involvement worth mentioning is the increasing emphasis on mandatory consultation with the workforce. This already exists with collective redundancies, business transfers, and health and safety arrangements. There are also moves toward consolidation into a general requirement for employers to establish works councils, largely on the German model. It is again important that such requirements are not, or would not be, restricted to that part of the workforce working on the employer's premises. For this reason, an employer of teleworkers would need to be aware of any legal obligations to involve teleworkers in the necessary consultations, though of course the machinery for doing so might be more complex.

Dismissal of Teleworkers

To a large extent, teleworkers would have no special position in relation to redundancy and other forms of dismissal. Once it becomes clear that they are employees, the normal laws (e.g., on length of notice and fair dismissal procedures) would apply to them. However, three particular possible complications exist. The first is that computer misuse could be a particularly difficult disciplinary matter when the employee is a teleworker. It is already causing problems within traditional office settings; for example, in relation to personal use during working times, the downloading of pornography or other offensive material, or the sending of offensive or defamatory e-mails. In the UK it was held, at an early stage in computerization, that almost any form of deliberate computer misuse during employment can potentially justify summary dismissal: (*Denco Ltd. v Joinson* [1991]—hacking into computer programs to which the employee had not been given access). However, the position could be more complex where the misuse was more 'recreational' or 'humorous,' and the company's approach had previously been lax. In the case of teleworkers, the boundaries between personal and business use could be even more difficult to lay down and police, possibly the more so if teleworkers use their own machinery. To what extent would the employer have to show tangible detriment to the business through the teleworker's activities in order to justify a dismissal? Potential problems with e-mail could be significant when it constituted the teleworker's principal and natural means of communication.

The second complication, in a sense following from the first, arises from the normal common law requirement that (except in a case of summary dismissal for gross misconduct) a dismissal must be with contractual and/or reasonable notice. This causes a problem with employees working with computers because of the possibility of employees under notice wreaking revenge by interfering with their employers' computer systems. In the normal situation of an office worker, this usually is countered in practice by giving pay in lieu of notice (so that the notice requirement is paid out and the employee is discharged immediately, before any damage is done). Where, however, the employee is a teleworker, this sort of physical neutralization would not be possible, and the employer would have to devise other ways of isolating the employee from the firm's systems, either electronically or by physical impounding of the machinery if it belongs to the company.

The third complication could arise where the firm faced a redundancy situation. If it were operating on any sort of basis that viewed office workers as the core staff and teleworkers as peripheral, there could well be a temptation to select the teleworkers first. However, this could be legally dangerous, either on ordinary grounds of unfair dismissal (why select only from one group if the work being done is similar?) or, more dangerous still, if there were evidence that the teleworkers were in practice predominantly of one gender, because it could then be argued that the method of selection constituted indirect sex discrimination. By analogy, the old UK habit of dismissing part-timers first has been potentially illegal for years now, an early casualty of EC law-led sex discrimination requirements. There are distinct advantages in the UK in pleading the claim as sex discrimination rather than unfair dismissal, because then there is no qualifying period of employment required and the damages obtainable are greater (*Clark v Ely [IMI] Kynoch Ltd., 1983*).

Drafting Teleworkers' Contracts

As with any other form of employment, the common law would not actually require a teleworker's contract of employment to be in writing. If necessary, the law will imply such a contract on an informal basis. However, the normal principle in employment law that a written contract is preferable would apply particularly strongly here, because of the desirability of covering some of the untypical aspects expressly. Also the traditional way of filling in gaps in informal contracts (the implied term) may not work well because it relies on a high level of impliedly shared expectations and assumptions, but with something as novel as teleworking, the parties may actually be starting with very different assumptions or desires. When drafting a written contract, in addition to the standard terms, the parties may need to give careful attention to matters such as the following:

- Working hours—can this be left flexible or is some quantification of 'normal' working hours necessary for the purpose of the state's or jurisdiction's minimum wage (in countries where this practice applies) or working time legislation? Must the employer get the teleworker's written agreement to exceed or disapply any normal legal requirements on matters such as maximum hours, rest breaks or night (and weekend) working?

- Payment issues—is payment to be on an hourly or work-done basis, or on a salary basis? If the latter, what is the expectation of the work to be done for it? Where, if at all, does overtime fit in, and how is it to be measured?

- Performance appraisal framework. With no direct observation a specific shift should be reflected in the system, for example, emphasis on results and outputs. This is of crucial importance when a performance-related pay system is applied in the organization.

- What records must the teleworker keep for pay calculation and legal purposes?

- Holiday entitlements—how do these fit in with a highly flexible regime, and will they satisfy any minimum legal requirements (e.g., the minimum four weeks paid holiday across the EU)?

- Sickness issues—when is the teleworker considered 'off sick' (an intrinsically illogical concept, given the set up), and what evidence is to be required?

- Personal completion of the work—is this to be a contractual requirement or could the employee use family or friends, or delegate?
- *Force majeure*—if work becomes impossible (e.g., electrical or electronic failure) do employees still qualify for pay or do they have to make the work up later?
- Ownership of electronic machinery—where does it lie?; who is liable for installation, insurance, maintenance and repair?; and what rights does the employer have (to hardware and software) on termination of employment?
- Liability insurance—who is liable to arrange and pay for public liability insurance for the teleworker?
- Employer participation in payments of office-related maintenance – who is to pay the phone and electricity bills?
- Confidentiality—in addition to normal common law implied terms, what does the employer want to specify in relation to the teleworker's access to and handling of sensitive information, including information on or belonging to customers?
- Disciplinary issues—what needs to clarified in relation to computer use and misuse, and what is likely to be the disciplinary result of misuse?
- Post-termination restrictions—if teleworkers have access to confidential information or trade secrets, and/or are likely to be subject to headhunting by other firms, does the employer need to specify a restraint of trade clause limiting competition for a period after termination of the employment? If so, what would be reasonable limitations for such a tie, in the light of the work that teleworkers have been doing, and the fact that they may have had much wider electronic contact with customers, suppliers etc., anywhere in the world than the traditional officer worker?

These major areas need to be covered under individual contracts as 'standard' contracts will not be applicable or sufficient. The above list can be used in developing contracts for new staff members destined to telework, and when employees who work in conventional terms transfer to teleworking terms. Changes to the legal system are slower than the pace modern business life imposes on employers. Thus it seems best for employers and employment solicitors to draft contracts that anticipate possible options and difficulties, and cover these exceptions under specific contracts designed for teleworkers, rather than leaving this to the court to decide.

CIVIL LIABILITY AND OCCUPATIONAL HEALTH AND SAFETY ISSUES

When civil liability and health and safety laws are applied to teleworkers, a similar picture to that in relation to employment law appears, namely, that the same laws will apply but with new adaptations and challenges. Again, these laws have evolved in the traditional context of work being done on the employer's premises, with fairly clear lines of control and discipline of employees. New challenges could arise in three main areas.

Civil Liability of the Employer Toward the Teleworker

In common law, an employer has a duty of reasonable care to an employee, which may be supplemented by statutory duties under the particular state's or jurisdiction's industrial safety legislation. Breach of either the general duty of care or the statutory duties may enable the employee to sue the employer for damages where an accident has been caused by that breach. This is trite law, and industrial accident cases usually account for about half of all personal injury litigation, along with road accidents and, increasingly, medical malpractice suits. However, the advent of teleworking could place new emphasis on two long-standing problems in this area of law and adds a new dimension to them.

The first problem is to delineate *which* accidents the employer may be liable for. Generally, employers only are potentially liable for work-related accidents, not for mishaps occurring away from work while the individual is acting in a 'private' capacity. In many simple cases the distinction is obvious (falling off a ladder at work, as opposed to falling off at ladder a home), but even in the case of traditional factory or office-based work the borderline has been notoriously unclear and has spawned much litigation. Just what is meant by the famous phrase 'in the course of employment'? To give a flavor of the problem, in one leading UK case, *Smith v Stages* (1989), the question was whether a car crash suffered by a peripatetic worker returning from a job in a fellow employee's car had occurred in the course of employment. The facts were neatly balanced since he was travelling back home (normally not included), but his time was being paid by the employer, and on balance, it was held that he still had been acting in the course of his employment. Similarly, there is a series of cases on whether sporting injuries (e.g., by a fireperson playing sport while on call, or by a policeperson playing for their force's representative teams) can constitute an industrial injury. This problem could be particularly acute in the case of the teleworker, where the parameters of the work itself are likely to be uncertain.

When are teleworkers 'at work'? What is the position if they are travelling for a purpose arguably related to the work, e.g., shopping for stationery required for the work (but also buying groceries)? What if they are attending a training course thought beneficial to their professional development (but not actually required by the employer)? What is their position during meal times? Questions such as these are crucial in preparing an injured teleworker for litigation. If decided in the teleworker's favor, that would not mean automatic success because the employer's obligation is only one of *reasonable* care. An exception to this would be if the plaintiff's action were for breach of statutory duty (rather than for common law negligence) and the relevant state's or jurisdiction's legislation on the point adopted strict liability on the employer, as a matter of policy.

Here again, teleworking presents problems, in particular the question of what can reasonably be expected of employers in the case where employees work at home in a relatively flexible manner—what level of training, advice and/or checking could be expected? An interesting parallel here is with employees who are seconded by a firm to work abroad for another company. At common law this does not absolve the firm of the normal employer-employee duty of care, though obviously the distance and separation are major factors influencing what the employer can reasonably be expected to do, e.g., by way of ensuring the safety of the employee's place of work: *Cook v Square D Ltd.* (1992) (an employer of a British worker sent to work in Saudi Arabia was not liable for injuries sustained when he fell over a floor tile at his place of work there). Could employers be expected, for example, to check the safety of materials, furniture, etc. provided by the employees themselves? Could the employer be liable for ergonomic problems incurred by employees while working long hours at a computer? Thus, the linked questions of what constitutes 'work' and how far the employer can reasonably be expected to take steps to advance the safety of that work could cause acute difficulties in the case of teleworking.

Moreover, the question of what constitutes 'work' could have a second significance for an injured teleworker in any state or jurisdiction that makes special welfare/social security provision for employees suffering work-related accidents or illnesses. These could include higher income-replacement benefits while off work or lump-sum benefits in the case of lasting impairment. Indeed, the well-worn phrase 'an accident arising out of and in the course of employment' originated in the earliest UK legislation providing for such benefits: the *Workman's Compensation Act, 1897,* which introduced the forerunner of the present system of industrial injury benefit. Thus, the genesis and categorization of the employee's injury or other incapacity can determine both entitlement to state benefits and the likelihood of a civil action for damages against the employer. The simple point being made here is that the distinctions that always have to be drawn could be even more difficult in the case of teleworkers.

The second problem concerns the types of injury for which the employer may be liable. The law of civil liability for personal injury grew up around cases of physical injury from one-off accidents, such as the amputation of an arm by an unfenced machine. Such a traditional accident could befall a teleworker (e.g., electrocution from faulty machinery), but that is much less likely to be the case than in a normal manufacturing industry. Instead, teleworkers may gain protection from two vitally important developments in civil liability law, generally, in the last thirty years. The first is that the law no longer is restricted to what might be termed the static condition of the workplace (Was that machine fenced? Was that handrail secure?), but now looks more broadly at whether safe systems of work have been provided. This can be a much more dynamic approach, involving questions of planning, supervision, training and general good management. If something goes seriously wrong for a teleworker, it is likely to be in the area of work organization. A good current example is the issue of repetitive strain injury (RSI) in people who use computer keyboards. This is not a traditional accident, but it may have much to do with the organization of the work, good keyboard practice and taking of rest breaks. The RSI example also demonstrates the second development, which is that the law no longer covers only definable, one-off accidents but now can also apply to illnesses or diseases that have longer onset periods, and (increasingly, in the light of modern medical research) be shown to be work-related. Lung diseases in miners forced the pace on this in the UK, leading to the development of laws that could then be used in other areas, such as the major litigation over asbestosis and mesothelioma.

One early and vital change in the law was to extend the UK limitation period (for bringing an action) from three years from the date the damage occurred to the present rule of three years from the date the damage occurred *or* from the date the person *realized* it had occurred. This is essential to cover insidious diseases with onset periods much longer than three years. Ailments that once were thought to be simply an unfortunate part of life in some industries (e.g., early deafness, asthma) may now be compensated by damages if the employer failed to take reasonable care to prevent them (once it is reasonably established and known that the occupational link exists). This development could be essential to teleworkers whose form of work might expose them to long-term risks of RSI, musculoskeletal problems or, if it eventually is proven, exposure to radiation from certain forms of electronic machinery such as mobile phones.

There is, however, one other form of non-accident liability that currently concerns employers and which could be a particular difficulty. This is liability for occupational stress, leading to definable psychiatric injury. This hit the UK headlines in one high-profile case, *Walker v Northumberland County Council* (1995), when a council was found liable for a social worker employee's second, debilitating nervous breakdown. This was caused by excessive pressure from a mounting caseload when no measures were taken to lessen it after his first breakdown. An appeal in the case eventually was discontinued and a settlement was reached for £150,000, a substantial amount in UK damages law. This case had a major effect on personnel and human resources professionals.

This situation has been a source of serious debate ever since, especially in forms of employment that have a long-hours culture. Of course, this example of a stress case reiterates the point made above that at common law there are no guarantees of safety, and the question is: What *reasonable* steps should the employer have taken? As before, this might be an easier question to resolve in the case of traditional staff working on premises, where signs of stress-related illness might be more easily detected in good time.

What is the employer of teleworkers reasonably expected to do? At one end of the spectrum, the employer who unreasonably insists on an excessive workload (in spite of complaints) could expect to be found liable, but at the other end, what would be the position of an employer who only provided the amount of work that most teleworkers could deal with comfortably? At what point and in what way would that employer be expected to realize that *this* teleworker was having serious problems coping? Employers may be tempted to think of teleworking as a way of isolating themselves from stress claims (how could I do anything—they were working at home), but if this area continues to evolve as it seems to be at present, so that employers are expected to have systems in place to combat workplace stress, the inherent uncertainties of teleworking could ultimately cost the employer.

Civil Liability of the Employer for the Teleworker

The other principal issue that might arise in civil liability law is when the employer might be liable in damages *for* the teleworker, i.e., to a third party injured by the teleworker's acts or negligence. In the case of a client who received poor workmanship or advice, or a poor product because of bad work by the teleworker, there may not be a problem because the client will have paid and therefore will have a breach of contract action directly against the employer. The status of the person completing the work is irrelevant. However, if a third party suffered harm independently inflicted by the teleworker (either physically, emotionally, or economically, e.g., by receiving poor advice directly from the teleworker), the question would arise whether the employer was vicariously liable for that harm. It is well established at common law that there can be liability for negligent misstatements leading to financial loss, especially where bad advice is given in a business or quasi-business context: *Hedley Byrne & Co. Ltd. v Heller & Partners Ltd.* (1964). For this reason, employers may wish to restrict teleworkers from direct contact with clients, if that is feasible, to restrict the possibility of such liability, for which the employer might then be vicariously liable (below). Once again, the cases on this are legion and the distinctions can be thin and difficult, even in the case of an ordinary worker working under direct supervision on the employer's premises.

For reasons already given, problems could be particularly acute in the case of teleworkers because of the flexible and discretionary nature of the work and their lack of supervision. Three further aspects of this point bear noting. The first is that liability also could arise because of accidents occurring to a third party in the teleworker's home, e.g., a visiting client, or a person repairing the electronic equipment, in particular that proportion of the home being used for the work. In that case, who would be liable? This could depend on the particular state's or jurisdiction's laws relating to what tends to be known as 'occupier's liability.' In the UK, it has been established that in more complex cases there can be more than one 'occupier,' so that both the employer and the teleworker could be liable (see, for example, *Wheat v Lacon* [1966]—for the case where a brewery and its public house manager both were held to be 'occupiers' so that the brewery could be sued by a visitor injured when falling down an unlit staircase). The second aspect is that, in light of uncertainties such as this, the employer and teleworker would be well advised to come to a clear agreement, when contracting, as to who will insure against this form of public liability. In cases of doubt as to where liability ultimately would lie, it may be advisable to include indemnity clauses to determine who eventually pays.

The third aspect is that the employer also could be liable vicariously for any unlawful discrimination perpetrated by the teleworker, e.g., any racial or sexual harassment by electronic means. The extent of such liability probably would depend on the laws of the particular state or jurisdiction and how strictly they put liability on the employer. In the UK, a particularly strong approach to this has been taken by the courts: The employer will be held liable for almost any discriminatory practices carried out by individuals while at work *unless* the employer can prove that it took all reasonable steps to stop that action, or actions like it, from happening (*Jones v Tower Boot Co. Ltd.*, [1997]). Anti-discriminatory policies, training and monitoring can be important here, but again these matters could be more difficult for employers of teleworkers.

Health and Safety Law

In many jurisdictions, industrial safety is not just left to the private law of civil liability. It is further strengthened by public law statutes that advance industrial health and safety, through health and safety inspectors, via administrative powers (e.g., to close down dangerous operations), and by imposing serious criminal penalties on employers for breach of these health and safety laws. Thus, in the case of an industrial accident, the employer could be sued for damages by the injured employee *and* prosecuted by the state authorities.

While the form of industrial safety legislation varies from country to country, it is being harmonized throughout the EU. The long-standing UK legislation (the *Health and Safety at Work Act*, 1974) is being used as a model, in particular in its wide coverage of all workplaces and proactive concern to prevent accidents and diseases, not just to compensate afterward. Health and safety law is now an area of EU competence, allowing the EU Commission to produce directives with the aim of harmonising health and safety laws throughout the Union: *Treaty of Rome*, Article 118A. In 1989 the Commission revolutionised this area with the *Health and Safety Framework Directive* (89/391/EEC) and five subsidiary directives: on the workplace, work equipment, protective equipment, manual handling and VDU equipment. These were adopted into UK law in 1992 by the *Management of Health and Safety at Work Regulations* and five subsidiary sets of regulations closely based on the directives (known colloquially as the 'Six Pack'). Together, they replaced all the old factories legislation dating back to 1802, and now form the backbone of modern UK law, along with the *Health and Safety at Work Act*, 1974.

Inasmuch as a teleworker is an 'employee,' these laws will apply, but also there may be specific applications. The general duties of the employer under the 1974 Act apply not only to employees, but also to the self employed (for health and safety purposes, it does not matter which category the teleworker is defined) and to anyone else affected by the employer's 'undertaking,' such as visitors to the teleworker's premises and others working there (e.g., on repairs or maintenance). In the landmark case of *R v Associated Octel Ltd.*, (1996), a chemical company was held criminally liable for an explosion causing injuries to an employee of another company performing cleaning and maintenance on its premises. This goes much further than the common law of civil liability.

It is clear under UK law that teleworkers' activities would be considered part of their employer's undertaking. Under the EU-inspired *Management of Health and Safety Regulations* (1992), there is a heavy emphasis on the compulsory carrying out of risk assessments (and their recording in writing) as a key means of accident prevention. Also, there have to be procedures for health surveillance of employees in light of any risks identifiable by the risk assessment. Again, it is clear that these regulations are broad enough to cover teleworkers, so in health and safety terms, it would be insufficient for employers simply to set up the arrangement contractually and then let new teleworkers go off and arrange it all for themselves. There is a positive obligation on the employer to assess the intended workplace and to repeat this assessment when necessary (e.g., on the introduction of new machinery).

More specifically, teleworkers' use of electronic machinery will be covered by the *Health and Safety (Display Screen Equipment) Regulations* (1992) which draw no distinction between places where such work is carried out. Again, there are requirements of risk assessments of the intended workstation that must meet certain specifications taken directly from the governing EU Directive and a general obligation on the employer to plan the employee's activities so as to ensure breaks or changes of activity to break up the work at the screen (also taken from the Directive are requirements of training and information and access to eye testing at reasonable intervals).

These requirements are bound to be more problematic for employers of teleworkers, but it is equally clear that under this statutory regime, doing nothing is not an option. Finally, as noted in the previous section on employment law, there is also EU law involvement in the modern requirements for consultation with the workforce on health and safety issues. This again comes from the *Framework Directive* and is enacted into UK law by the *Health and Safety (Consultation with Employees) Regulations* (1996). As teleworking is likely to be an area of little or no trade union involvement, this means adopting the alternative method of establishing consultative machinery with directly elected employee representatives. In such an exercise, teleworkers would *not* be excluded and would have to come within one or the other appropriate 'constituencies.' The bottom line is that health and safety law is one area where it is particularly important that employers do not assume they can just contract with teleworkers and expect to have nothing more to do with them than to receive the work and sign their checks periodically.

OTHER CONSIDERATIONS TO BE TAKEN

Whereas many of the legal aspects concerning Teleworking are covered by both employment contracts and Health and Safety issues, there are other elements that should be of great concern and have legal aspects to be aware of. These can be divided into two categories: 'Hard' and 'Soft'. By *Hard* we mean elements that are (relatively easily) measurable, most notably the issue of costs. By *Soft* we mean other aspects concerned with teleworking where it is difficult to point out the exact amount of money or other resources, the intangible aspect, but that are enormously important for the employee (and sometimes for the employer—see reputation). Among these are career issues, discrimination, confidentiality and others.

Hard issues easily can be related to costs. Cost of travel; direct bills, such as telephone; as well as indirect ones: electricity/gas (for heating), maintenance, different levels of insurance costs, etc. In particular, the cost of the home office should be considered at each stage of teleworking: from pioneering, temporary, on-the-spot development to an established practice. In the pioneering stage the employer may suggest that the teleworkers take home their old PC, because the company has just bought upgraded equipment for the office. The employer may expect the employees to set up the equipment using, for example, their own cables, old desks and shelves, and the rest of the facilities.

In the established stage, employers may need to buy equipment compatible with that in the office and have the company's IT department install it (only after an inspection is carried out for safety aspects). The at-home workstation will be furnished with company shelves, desk, and the rest – there needs to be an agreed 'package' of supplies for teleworkers.

Associated with costs is working time. Employees who commute to and from work save a considerable amount of time and travel costs if they transfer into teleworking. These gains cannot be counted as a good reason for the employer to expect teleworkers to spend more time on work on a formal base (but this may be expected through an informal agreement). There is also the possibility of involuntary breaks such as power outages, e-mail system failures and other *force majeure*-related causes. If these happened at work, employees would get paid for the time spent at work. It would not matter if they missed two hours of work because of a fire-drill or if there were a power outage. What is to be done with employees who cannot work at home because of similar problems? Is it a time out-of-work or a part of work not used? Employees may be able to claim compensation for these situations under local legislation (e.g., the *US Fair Labor Standards Act*).

Quite a different case occurs when employees cannot afford to or are strongly reluctant to transfer to teleworking say, because of a small house. When can employers still expect them to transfer to this mode of work? Can employees expect financial support to build a new annex to their houses for work purposes? What are the taxation implications for such added value to the building? What about the 'soft' element, such as the case of a family that is reluctant to accommodate such considerations? Can employers force teleworking on reluctant employees? The case would be different, of course, if a new teleworking job were offered, but when an existing 'conventional' job is altered into teleworking, considerations such as resistance should be considered.

When a large insurer decided to close its offices, all employees in certain departments were told they had to operate from home as no space had been designated for them. One employee, an elderly person living in a small one-bedroom house, found he needed to work in his kitchen (Baruch and Nicholson, 1997). This may mean a redundancy de facto, since it forces the employee to leave the organization. Thus, if they refuse to telework on such a strong ground, employees probably will be entitled to redundancy payments.

Other *Soft* issues are Career advancement, recruitment and indirect aspects. Career prospects may be different for teleworkers, particularly when only some employees are teleworkers. Under such circumstances, teleworking might be an impetus for claims against poor career opportunities for teleworkers. Baruch and Nicholson (1997) found that teleworkers claim to be excluded (out of sight is out of mind when promotion is under consideration was the general opinion in their study). Thus suggesting, or even forcing employees to telework might mean indirect discrimination in terms of career advancement. If women or disabled personnel are especially encouraged to telework, this may mean an indirect sexual /disability discrimination. As Clarke (1999) indicated, being active and involved in 'office politics' can be a way of developing networks and reaching an influential position within the organization. Working from home, particularly on a full-time base, isolates people from these networks and reduces their visibility and chances of development.

Indirect discrimination against deprived populations can be associated with teleworking. If a company only offers teleworking positions, this might exclude people who live in poverty and cannot afford a spare room, which is essential for effective teleworking. By practicing only teleworking options for employment, employers might face future claims of indirect discrimination.

In relation to intellectual properties, in today's knowledge-based industry, much of the knowledge dwells within the individual. How can we distinguish between software developed by employees during their working hours and software developed on their free time? Sometimes the standard copyright law can be applied directly, while in other cases it must be interwoven into the contract.

Business confidentiality may suffer greater risk in a home office. There is a potential for loss of sensitive or proprietary data stored on home-based computers if employees chooses to open corporate networks and company information to people in the public realm. Employees also could introduce computer viruses into corporate networks by using their own equipment. This could go also in the opposite direction, from company networks to individuals' equipment, possibly harming peoples' intellectual property stored in their private files. While such possibilities can be controlled by making sure sensitive information is partitioned on an employee's hard drive and that passwords are required, this area will be an ongoing concern to the company and should be addressed up front.

We conclude the *soft* issues by looking at the ethical aspect of teleworking. Moon and Stanworth (1997) suggest that for ethical reasons, teleworkers should be on a voluntary base; be able to return to on-site work; receive equal pay and benefits; be appraised in the PA system; get allowances for costs; be provided with the needed equipment; have opportunities to meet together; be sent to training and development classes similar to 'conventional' employees; be considered for promotions; and, when relevant, have access to trade union information and membership.

This normative approach of what organizations *should* apply, sometimes conflicts with the realities of practical management. For example, the idea that teleworkers should be able to choose to return to work on-site can be irrelevant and impractical if the offices are closed, especially in cases when the impetus for teleworking is to save office space.

CONCLUSIONS

Conclusions and implications will be discussed separately at national, employer and individual levels. The legal aspects of teleworking have been neglected in the literature, probably for two reasons. On the management side, from a lack of awareness and having to deal first with more practical issues. On the legal system side, from the long time that typically characterizes the legal system's response in most countries to rapid environmental changes. Any of these problems imply that employers and, in particular, legal advisers should pay considerable attention to facts and circumstances that result from teleworking, and incorporate them into updated employment contracts. Clearly, the forms of legal enforcement that could be used in the case of infringements of a teleworker's rights will vary from jurisdiction to jurisdiction and may involve criminal penalties as well as civil damages; however, by applying a proactive approach, employers can avoid future embarrassment in court. It would be much better for all involved if terms and conditions were set in advance, as clearly as possible. For example, with regard to the crucial issue of 'what is work,' the contract can define this.

National Level

Organizational policies and practices would not release national legal systems from their duty of developing a legislative system of acts and regulations to cope properly with the variety of issues concerned with teleworking, many of which were discussed in this chapter. First, basic laws should be adapted to deal with questions specific to teleworking. However, teleworking is potentially so different that, at least in certain areas, a change of the legal regime is called for.

Within the EU during the last two years specific directives were passed and applied concerning part time employment and fixed-term temporary workers. These covered not only the banning of discrimination against such workers but also obligations to improve their conditions and treat them more as a valued and integral part of the workforce (e.g., how to move from one status to another). Perhaps it is now the time to consider a public regulatory system along similar lines for teleworking.

Until this is done, we would encourage employers and employees to be proactive, to recognize the positives and possible pitfalls of teleworking, and to deal with the legal aspects of this mode of work.

Organizational Level

Organizations should refer to teleworking at three levels: philosophy, policy and practice. At the philosophical and cultural levels, it is a question of how teleworking fits in the present or anticipated future direction. If a fit exists, the next stage is to develop a policy that accommodates teleworking in a way that will produce effectiveness and efficiency while avoiding legal problems. The practice level then follows by establishing a set of 'best practices' on both the operational level of teleworking and the legal level of proper employment contracts. These will help employers avoid potential claims (e.g., resulting from the *Fair Labor Standards Act* in the US or EU legislation in Europe). Companies and organizations should develop clear guidelines applicable to any employee working at home, and should give specific attention to the hours when work may be performed and the total hours of work expected per week.

Individual Level

Individuals should be proactive, understand that teleworking is a relatively new mode of work and before opting to work at home, investigate whether their individual and home situation can support teleworking. They should be careful when agreeing to new contracts and make certain, for example, that the contract permits a reversal of the transfer to teleworking if it proves to be impractical. There also are differences between self-employed, subcontractor and teleworker employees. At a separate level there is a need to distinguish among them. What are the benefits and possible pitfalls for each and what issues should be taken into account when signing a new contract or changing an old contract into a part- or full-time teleworking arrangements?

Final Note

We hope our readers will have obtained useful guidance and advice from this analysis, whether as employee or employer, with regard to employment contracts, health and safety matters, and the more general *hard* and *soft* issues outlined above. If it sometimes appeared that we have been more concerned with raising questions than providing specific answers, it is because to date many of these questions have not been properly asked. Much of this area deserves further exploration, inclusion of what precedents exist, and lateral thinking about the problems that may arise. At a minimum, what we hope we have established is that doing nothing no longer is an option.

REFERENCES

Airfix Footwear Ltd. v Cope [1978] ICR 1210, EAT.

Arthur, M. B., Inkson, K. and Pringle, J. K. (1999). *The New Careers*. London: Sage.

Arthur, M. B. and Rousseau, D. M. (1996). *The Boundaryless Career*. New York: Oxford University Press.

Baruch, Y. and Nicholson, N. (1997). Home, sweet work, *Journal of General Management*. 23(2), 15-30.

Beatson (1995). Progress towards a flexible labour market, *Carmichael v National Power* [2000] IRLR 43, HL

Chesbrough, H. W., and Teece, D. J. (1996). When Virtual Virtuous? Organizing for Innovation, *Harvard Business Review*, Jan-Feb, 65-73.

Clarke, J. (1999). *Office Politics*. London: The Industrial Society.

Clark v Ely (IMI) Kynoch Ltd. [1983] ICR 165, EAT.

Cook v Square D Ltd. [1992] ICR 262, CA

Cross and Raizman (1986). *Telecommuting: The Future Technology of Work*. Homewood, IL: Dow Jones-Irwin.

Davidow, W. H., and Malone, M. S. (1992). *The Virtual Organization*. New York: Harper-Collins.

Denco Ltd. v Joinson [1991] ICR 172, EAT

EC (1994), *Legal, Organisational and Management Issues in Telework*. (Bruxelle: European Commission).

Employment Gazette 55; 88, 225

Felstead, A. (1996). Homeworking in Britain: the national picture in the mid-1990s. *Industrial Relations Journal*, 27, 3, 225-238.

Grant, K. A. (1985). How Practical is Teleworking?, *Canadian Datasystems*, 17(8), 25.

Gray, M., Hodson, N., and Gordon, G. (1993) *Teleworking Explained*. Chichester: Wiley.

Handy, C. (1995). Trust and the Virtual Organization, *Harvard Business Review*, 73/3, 40-50.

Handy, S. L. and Mokhtarian, P. L. (1996). The Future of Telecommuting, *Futures*, 28(3), 227-240.

Hedley Byrne & Co Ltd. v Heller & Partners Ltd. [1964] AC 465, HL.

Hellyer Bros Ltd. v McLeod [1987] ICR526, CA.

Huws, U. (1993). *Teleworking in Britain*. The Employment Department Research Series 18.

Huws, U., Korte, W. B., and Robinson, S. (1990). *Telework: Towards the Elusive Office*, Chichester, UK: Wiley & Sons.

IDS (1996). IDS Study 616, (London: Income Data Services).

IRS (1996). Teleworking in Europe: Part One, *European Industrial Relations Review*, 268, 17-20.

Jones v Tower Boot Co Ltd. [1997] ICR 254, CA.

Jones, J. C. (1957, 58). Automation and Design parts 1-5, in *Design*, 103, 104, 106, 108, 110.

Kelly, M. M., (1985). The Next Workplace Revolution: Telecommuting, *Supervisory Management*, 30/10, 2-7.

Korte, W. B., and Wynne, R. (1996). *Telework*, Amsterdam: IOS Press.

Lamond, D., Daniels, K., and Standen, P. (1997). Defining Telework: What is it Exactly?, *Second International Workshop on Telework Amsterdam 97*.

Mitchell, W. J. (1995). *City of Bits*. Cambridge, MA: MIT Press.

Moon, C. and Stanworth, C. (1997). Ethical issues of teleworking. *Business Ethics: A European Review*, 6(1), 35-45.

Nethermere (St Neots) Ltd. v Gardiner [1984] ICR 612

Negroponte, N. (1995). *Being Digital*. London: Fontana.

Nilles, J. M., Carlson, F. R., Gray, P., and Hanneman, G. J. (1976). *The Telecommunications-Transportation Trade-Off*, Chichester, UK: John Wiley and Sons.

Nolan, N., and Walsh, R. (1995). The structure of the economy and labour market, in Edwards (Ed.), *Industrial Relations, Theory and Practice in Britain*.

Olson, M. H. (1988). Organizational Barriers to Teleworking. In Korte, W. B., Robinson, S. and Steinle, W. J. (Eds.), *Telework: Present Situation and Future Development of a New Form of Work Organization*. Amsterdam: North-Holland.

Peiperl, M. A. and Baruch, Y. (1997). Back to Square Zero: the Post - Corporate Career, *Organizational Dynamics*, 25(4), 7-22.

Qvortrup, L. (1998). From teleworking to networking: definitions and trends, in P. J. Jackson and J. M. van der Wielen (Eds.), *Teleworking: International Perspectives* (London: Routledge), 21-39.

R v Associated Octel Ltd. [1996] 4 All ER 846, HL.

Shamir, B. and Salomon, I. (1985). Work-at-Home and the Quality of Working Life, *Academy of Management Review*, 10(3), 455-464.

Smith, I. T. and Wood, J. C. (1996). *Industrial Law*. 6th edition. London: Butterworths.

Smith v Stages [1989] AC 928, HL.

Stanworth, J. and Stanworth, C. (1989). Hometruths about teleworking, *Personnel Management*, November, 48-52.

Toffler, A. (1980). *The Third Wave*, London: Collins.

Trodd, E. (1994). What is Teleworking? *British Journal of Administrative Management*, Dec 1993/Jan 1994, 9.

UK White Paper *People, Jobs and Opportunity,* (Cm 1810, 1992).

Van der Wielen, J. M. M., and Taillieu, T. C. (1995). Recent Conceptual Developments in Telework Research. *Proceedings of the Organizational Management Group*, 13/2, 19-26.

Walker v Northumberland County Council [1995] IRLR 35.

Wareing (1992). Working Arrangements and Patterns of Working Hours in Britain. *Employment Gazette,* 100(3).

Wheat v Lacon [1966] AC 552, HL.

ENDNOTES

[1] In a much-quoted dictum, the nineteenth century jurist Sir Henry Maine said that the movement of the progressive societies has been "from status to contract."

[2] This definition (taken from the sex and race discrimination legislation, where it has always applied) has been used in the *Working Time Regulations* 1998, the *Minimum Wage Act* 1998 and the *Public Interest Disclosure Act* 1999. There is a power in the *Employment Relations* Act 1999 for the government to extent it to *all* other existing employment laws. Interestingly, a recent decision of the Court of Appeal has stressed the particular importance of personal service as one of the factors pointing to a contractual relationship of employment under the common law tests: *Express and Echo Publications Ltd. v Tanton* [1999] IRLR 367.

Chapter II

Teleworking in Ireland: Issues and Perspectives

Frédéric Adam and Gregory Crossan
University College Cork, Ireland

ABSTRACT

Teleworking has been used since the early 1970s in some countries, but it is still relatively underdeveloped in Ireland. This study sought to establish why this is and concentrated on the implementation of teleworking arrangements from both management and teleworker perspectives in Irish organizations. This study indicates that, in the majority of cases where teleworking exists, it has been implemented in an ad hoc manner and is largely employee-driven. Teleworking is not actively encouraged and top management commitment does not exist. It seems Irish managers are not yet persuaded of the benefits inherent in the concept of telework or that they are uncertain whether the benefits are worth the risks resulting from the introduction of this new method of organizing work. This is unfortunate given the very positive experiences with teleworking reported in this study.

INTRODUCTION

Although teleworking has been in operation in the United States since the oil crisis in the early 1970s, it is still a relatively new and underdeveloped method of working in Ireland (Bertin & O'Neill, 1996). This is surprising given the advanced state of the telecommunication infrastructure in this country and the high uptake rate of the Internet/World Wide Web by both the Irish public and organizations (Yahoo!, 1998). This study of the issue of teleworking found little Irish specific empirical research into the practical implementation of teleworking in organizations and into the barriers to adoption that managers in organizations may be facing. Another question that seemed under-researched is whether teleworking offers opportunities for greater integration of the disabled in the workforce and whether these opportunities are being seized.

In the absence of clear, recent empirical evidence, the purpose of this research was to conduct an exploratory study into the implementation of teleworking arrangements from both management and teleworker perspectives. The research objectives were focused on the type of teleworking implemented, the policies and procedures implemented to manage teleworking, the type of work and individual best suited to telework, the technologies used, the driving forces and obstacles and, finally, whether teleworking can be used to integrate people with disabilities into the workforce. A combination of data collection methods (secondary data from literature, survey of managers and workers involved in teleworking and follow-up telephone interviews with both managers and teleworkers) was used in the research strategy in order to strengthen the findings of the research by triangulation.

The results of the study indicate that, in the majority of cases where teleworking exists, it has been implemented in an ad hoc manner and is largely employee-driven. Generally, teleworking is not actively encouraged and top management commitment does not exist. As a result, adequate IT and communications support is not forthcoming as teleworking is not seen as a priority by IT management. Finally, it seems that teleworking is not a panacea to the employment ills of all disabled people, but where it is suitable it could be encouraged with great benefits. In this regard, it seems that more could be done to try to integrate disabled people into the workforce in a meaningful way given the high potential of teleworking for this purpose.

Overall, our research indicates that Irish managers are not yet persuaded of the great benefits inherent in the concept of telework or, at least, that they feel uncertain whether the benefits are worth the risks resulting from the introduction of this new method of organizing work. This is unfortunate given the very positive experiences with teleworking reported by both workers and managers who took part in this study.

DEFINING TELEWORKING

Nilles proposed the concept of teleworking during the first international oil crisis (Nilles, 1994). He focused critically on the waste of energy in private and public transport systems and suggested that the science of information technology had the potential to substitute electronic communication for physical travel (Chapman, Sheehy, Heywood, Dooley & Collins, 1995; Reid, 1993). Soon after, the concept of meshing telecommunications and personal computers emerged with the term *telematique* first coined by Nora and Minc in 1978 (Berry, 1996).

Diversity is the most striking characteristic of available definitions of teleworking, which can be found in abundance. However, an analysis of existing definitions reveals that more than 60 per cent of definitions are based on the combination of two or three core concepts: organization, location and technology (de Beer & Blanc, 1985). It therefore seems appropriate to define telework as a form of organization where:

> Work is performed in a location remote from central offices or production facilities, thus separating the worker from personal contact with co-workers there. New technology enables this separation by facilitating communication.

Within this broad definition, two main types of telework can be distinguished:

> • Telework performed in a location near or in the worker's home, which responds to the needs of disabled or home-bound workers and helps to reduce commuting.
> • Telework performed in a business-determined location, which is primarily aimed at cost reduction or at servicing the market better.

Offshore work, whereby workers based in one country are employed by organizations based in other countries to reduce their data entry and data processing costs, is a typical example of the second type of telework that is commonly used in Ireland. For instance, New York Life employs workers based in Castleisland (in the west of Ireland) to enter claim forms while their American colleagues sleep, which saves the company at least 25% in salaries and associated costs. Such arrangements are not just beneficial to the company because the workers employed would have been unlikely to emigrate and are therefore given opportunities that would never have been available to them otherwise. Also, it is less damaging to the social fabric of economically less-developed areas. This type of work has developed in

Ireland more than in other countries because it is seen as a way to develop poorer regions where spare computer-literate manpower is available (Ettighoffer, 1992).

Since telework is carried out at a distance from the employer, communications technology must be used to link the two parties. Bertin and O'Neill (1996) stated that teleworking involves using phones, faxes, computers and other technologies to keep in contact. It is therefore the use of Information and Communication Technologies (ICTs) that distinguishes teleworkers from traditional homeworkers (Incomes Data Services Ltd., 1994). In this respect Ireland is quite well equipped for teleworking. The telecommunication infrastructure is well developed thanks to the policies pursued by the Industrial Development Authority (IDA) to attract multinational organizations to Ireland. Indeed, these organizations often put the availability of advanced telecommunication infrastructures as a condition for their choice of a location. Thus, Ireland is well prepared for teleworking and, according to the Yahoo! server (1998), Ireland is in the Top 10 countries in the world for Internet / WWW development.

JOBS SUITABLE FOR TELEWORKING

The information revolution has brought with it the requirement for a new type of work—one that depends more on the intellectual processing of information than on physical labor. People who do this type of work are often referred to as *knowledge workers* (Kling & Sacchi, 1982). They normally do not need access to industrial equipment or raw materials resources to do their work, but they need access to the flows of raw data from which they create information. Thus, as long as knowledge workers have access to the required information and IT infrastructure, they need not be bound to any one physical worksite (Kling & Sacchi, 1982).

The occupations for which telework is suited are primarily those that extensively use office technologies. According to previous research these can be divided into two major groups: clerical and secretarial occupations on the one hand, and technical and managerial on the other (Carruthers, Humphreys & Sandhu, 1992). Work that already is centered on the computer and telephone is a prime candidate for telework (Rogerson & Fairweather, 1997). Thus, IT specialists such as software programmers and systems engineers might be able to telework, but professional and management specialists such as financial analysts, accountants, public relations staff, graphic designers, and general managers are also potential candidates. Similarly, support workers such as translators, bookkeepers, proofreaders, draughtsmen, researchers, data entry staff, telesales staff and word processor operators could all telework, given suitable circumstances and arrangements (Rogerson & Fairweather, 1997).

Computer manufacturing companies have expressed a particular interest in developing telework and being competitive in the application of computer and software systems. International Computers Limited (ICL) in the United Kingdom introduced telework as far back as 1969 to retain skilled staff. IBM in the United States made telework options available in 1988 to employees on leave. NEC, a Japanese Computer company, established a satellite office in a suburb of Tokyo in 1984 in a bid to overcome a shortage of software development engineers (Carruthers, Humphreys & Sandhu, 1992). Thus, jobs suitable for telework should: (1) involve a high degree of intellectual, rather than manual work; (2) involve a certain degree of initiative; and (3) have measurable outputs (Stanworth & Stanworth, 1991a).

KEY MANAGERIAL ASPECTS OF TELEWORKING—COMMUNICATION AND COORDINATION WITH TELEWORKING

Some authors suggest that managerial styles may not need much transformation to cope with teleworking (Burch, 1991), while others argue that whole new managerial techniques must be developed (Reid, 1993). A major disadvantage of teleworking, in the minds of employers, is that it is perceived to be hard to manage (Huws, 1996a). Indeed, recent research has shown that coping with day-to-day management is the single greatest source of anxiety for managers who are considering the introduction of teleworking (Huws, 1996a).

A major problem faced by managers in relation to the implementation of teleworking is disruption to the communication channels in the conventional office (Kugelmass, 1995). In the telecommuting environment, opportunities for interaction are reduced (Fritz, Narasimhan, & Rhee, 1998). A number of authors have speculated that changes in the nature of interaction caused by telecommuting will negatively influence the performance of work activities (Huws, Korte & Robinson, 1990; Kraut, 1994). In an extensive study by Huws, Korte, and Robinson (1990), participants described the lack of social interaction with colleagues as a disadvantage of remote work, but other authors maintain that, proper management of the relationship between managers and teleworkers can increase the overall satisfaction with communication. More formalized communication channels such as scheduled meetings can be implemented (Kugelmass, 1995), and the use of ICTs, such as e-mail, for communication and information gathering can be encouraged (Kugelmass, 1995; Kraut, 1994).

Thus, when coordination is necessary, it can take place through different communication modes. The use of electronic coordination methods such as e-mail, electronic bulletin boards, or electronic project tracking tools is particularly useful because it enables asynchronous types of interaction whereby individuals participate as their schedule permits (Kraut, 1987).

EMPLOYEE SELECTION AND TRAINING

It is also important to identify people who have the right qualities for teleworking. Gray (1993) suggests teleworkers should be self-motivated, capable of working with minimum supervision, able to cope with minimal social contact, safety conscious, well organized and good communicators. Huws (1993) added that they should be mature, trustworthy, and self-disciplined. However, employers may have greater confidence about trustworthiness and loyalty if the individuals have worked for the organization for some time and their reliability and performance already have been proven (Murray, 1995; Chapman, Sheehy, Heywood, Dooley & Collins, 1995).

Much of the training for teleworkers will relate to breaking conventions associated with the office. Computers and videos could be used effectively with distance learning, but should be supplemented by on-site training in the form of lectures, group discussions and case studies (Stanworth & Stanworth, 1991a). It is recommended that new teleworkers receive job-related training, generic training and training in self-management (Huws, 1996a).

SUPERVISION AND CONTROL

Some managers feel concerned about the perceived loss of control when employees cannot be supervised directly (Meehan, 1997). This is the biggest concern in the day-to-day management of teleworkers. According to Hewitt (1993), the *working-time culture*, which values attendance, pervades most organizations. Nilles (1994) stated the perception that there is no way to know if an information worker actually is working, is prevalent amongst managers. Burch (1991) suggested that management focus should be on monitoring output and Huws (1996a) has suggested a variety of methods for monitoring telework:

1. Progress meetings between teleworkers and their managers.
2. Regular phone conversations between teleworkers and their managers.
3. Self-management because many teleworkers are professional and managerial staff and, therefore, are largely self-managing.
4. Setting targets can be an effective method provided the targets are precise enough to be monitored properly.
5. Analysis of output.
6. Payment by results.
7. Team meetings can substitute for the regular progress meeting with the manager, when teleworkers work as part of a team.

In any case, teleworking is likely to have significant implications for the management of people.

INTEGRATING THE DISABLED
IN THE WORKFORCE

A wide range of assistive devices now are available to allow people with special needs to obtain productivity levels comparable with other workers (Sandhu & Richardson, 1989). Many of these have been developed specifically to enable people with disabilities to compete for employment (Carruthers, Humphreys & Sandhu, 1992). In addition, many governments have introduced legislation to encourage employers to make the necessary changes in accommodation and work practices to facilitate disabled workers. Teleworking allows the disabled to work in their own customized environment and to interact with employers in the same way as able-bodied teleworkers (Gray, 1993). Thus, it could become a significant factor in helping companies employ a more acceptable percentage of disabled employees (British Telecom, 1992).

One of the greatest challenges for disabled people in the traditional office job is the trip to work. Clearly, full-time telework overcomes this obstacle and even part-time telework can provide considerable benefit. Working at home lets people operate in a physical environment tailored to their needs. A further benefit is the flexibility to work hours that suit the individual, which is particularly important for those who tire quickly and need regular breaks.

A basic tool for any teleworker is a PC or data terminal. One of the major concerns, therefore, will be how disabled teleworkers interact with their computer. Fortunately, a number of suitable user interfaces and input devices exist. The declining cost of high-resolution large screen PCs makes it possible to tailor the display of text by providing large fonts for the partially sighted. The use of graphical interfaces makes it easier for people with mobility problems who find typing difficult. Input devices such as mice, track balls, light pens and touch-sensitive screens, as well as special keyboards, provide some degree of tailoring for disabilities. The most recent advances in voice interaction now provide further opportunities to work, particularly for the visually impaired and those with severe arthritis.

More widespread adoption of teleworking, coupled with developing information and communication technology and services, undoubtedly will assist the disabled in securing employment using information technology (Gray, 1993).

CONSIDERATIONS FOR THE ORGANIZATION

A key organizational issue raised by teleworking is measuring and defining productivity. Teleworkers have been found to be between 30 and 100 percent more productive than their counterparts in the main office (Reid, 1993). Possible reasons may include removing the stress of commuting, more flexible working hours and teleworkers' appreciation of the greater responsibilities that come with teleworking (Huws, Korte & Robinson, 1990; Burch, 1991).

Reduced overhead is another major organizational benefit (Fowler, 1996), especially through reductions in total office space required (Littlefield, 1996). Efforts to reduce office space also have led to the emergence of the concept of *hot-desking*: a workstation area that is used by different people at different times (Huws, 1996a).

Teleworking also helps organizations retain rare skills; key employees who have to move still can keep their jobs (Rogerson & Fairweather, 1997) —or in solving skill shortages in certain locations (Ettighoffer, 1992). In the case of disabilities, teleworking could include those situations where existing full-time staff have an accident or develop a deteriorating health condition. Teleworking also allows organizations to penetrate nontraditional labor market sources, such as people with young children or elderly dependents.

But one of the main advantages of teleworking is the flexibility it provides. People can work in geographically dispersed teams that can be assembled and reassembled as the needs of the enterprise change. Team telework emphasizes the use of groupware technologies to support geographically dispersed information workers working on common tasks. This shift from the notion of electronic homeworking and telecommuting toward new methods of organizing work can significantly improve the applicability, credibility and acceptability of telework in the real world (Li, 1998).

Other benefits of teleworking include motivation, because employees respond positively to the trust and confidence indicated by its adoption (Chapman, Sheehy, Heywood, Dooley & Collins, 1995) and loyalty, as teleworkers are more loyal to the organization than their on-site colleagues (Huws, 1996a). Conversely, a weakening corporate culture (IRS, 1996), difficulties inherent in remote management and supervision (Reid, 1993) and additional burdens on communication (Rothwell, 1987) may become problems for managers.

CONSIDERATIONS FOR THE INDIVIDUAL

Huws, Korte and Robinson's (1990) survey found high satisfaction among teleworkers with all aspects of their working life. Uninterrupted concentration time in performing work contributes to job satisfaction (Benjamin, 1996). Stanworth and Stanworth (1991a) also found freedom from time constraints is a considerable advantage, in that individuals can work whatever hours suit them as long as the work is completed on time. Furthermore, teleworkers can lower their living costs by substituting electronic communication for physical commute, thereby reducing their travel costs, stress and time (Weiss, 1998; Goll, Lilienthal & Zapp, 1998; British Telecom, 1992).

However, studies of teleworkers found many respondents cited social isolation as the most serious disadvantage of teleworking (Huws, 1984; Rogerson & Fairweather, 1997; Goll, Lilienthal & Zapp, 1998; British Telecom, 1992). Disabled teleworkers often cite a desire for social contact as a reason for seeking work, a need less easily met by teleworking. Also, remote workers may find themselves without such on-site support systems as clerical backup, and there is a danger that the responsibility for routine tasks of this kind is left to the individual teleworker (Stanworth & Stanworth, 1991b; Hesse, 1995). Finally, teleworkers may feel obliged to work longer hours and produce more than their colleagues in the office do. This tendency is often exacerbated since there are no clear demarcations between work and personal time in the home (Hesse, 1995).

BARRIERS TO THE GROWTH AND DEVELOPMENT OF TELEWORKING

Despite its potential, telework has not spread as rapidly as predicted by the optimistic forecasts of the early 1980s. Technological constraints have played a role in this. Reduced electronic equipment cost have not been matched by reduced telecommunication services costs. Telecommunication charges still are distance related, which penalizes the diffusion of telework and existing networks are not always compatible, so their integration is difficult to achieve (Carruthers, Humphreys & Sandhu, 1992).

Nevertheless, the major impediments to more rapid and extensive adoption of telework seem to be organizational and cultural. Obstacles to the use of telework by employees with disabilities frequently are managerial rather than technological (Olson, 1987; Kraut, 1989; Semler, 1989). Technology that provides most of the requirements already is available, but the high cost of equipment and communications makes it difficult to demonstrate clear monetary cost savings to businesses. In addition, resistance to teleworking is not the sole prerogative of management. IT staff may resist the idea because they consider the cost of installation, support and maintenance of telecommuter equipment to be unacceptably high and difficult to manage (Warwick, 1996).

TELEWORK IN IRELAND TODAY

According to previous research, 1.4% of the Irish workforce could be regarded as home teleworkers (Bertin & O'Neill, 1996) and about 100,000 Irish people work from home at least on an ad hoc basis, including 15,000 full-time teleworkers (McDonald, 1996; Meehan 1997). However, these figures must be considered with caution because they depend upon the definition of teleworking being applied (Huws, 1994). A report produced by the Henley Center found that at least 10% of Irish workers would like to experiment with teleworking (Meehan, 1997). However, several barriers must be overcome if such growth is to be achieved. Key barriers are the cost of information and communication technology and the conservative nature of many organizations (Parliamentary Office of Science and Technology, 1995). Lack of competitively priced telecommunications services has been identified as a major weakness that could inhibit Ireland's progress in achieving the full benefits of the information society (Information Society Ireland, 1996). Ireland has one residential voice telephony provider, Eircom, and two mobile operators, Eircom's Eircell and Esat Digifone. Mobile infrastructure still is patchy in many rural areas. However, major developments in the telecommunication infrastructure were due in 1999, especially in terms of ISDN service and they may improve the situation.

SURVEY OF TELEWORK

This study aimed at carrying out an exploratory study of teleworking in Ireland both from the perspective of organizations and of individuals involved in teleworking. The research strategy involved a combination of data collection methods based on (1) an examination of existing research, (2) postal and e-mail surveys of individuals and managers involved in teleworking, and (3) follow-up interviews aimed at collecting complementary data. Secondary data was obtained from a review of the literature including books, journals, periodicals, conferences, newspapers, government publications and European Union publications. Information also was gathered at a conference on Teleworking, "Disability and Teleworking: Opportunity or Isolation," held in the European Foundation for the Improvement of Living and Working Conditions, Loughlinstown, Co. Dublin, in November 1998.

To identify organizations that could respond to the questionnaire, a telephone survey was undertaken of organizations in the IT and telecommunications sectors as identified in the *Top 1000 Companies in Ireland* in the February listing of *Business and Finance* (an Irish magazine that sells corporate listings). A list of names also was obtained from the National Rehabilitation Board of companies that had been awarded the NRB's Positive Towards Disability accreditation. Contacts were secured in 28 companies that had experience of teleworking and that were willing to respond to the questionnaire (in the end, only 14 actually returned questionnaires could be used).

A separate questionnaire was designed to gather information from the teleworkers' perspective. *Telework Ireland* was contacted and agreed to distribute the questionnaire via e-mail to all of its members. Some questionnaires also were distributed by post as contacts were made during the course of the research. The respondents covered the spectrums of the self-employed and those employed by an organization, able bodied as well as disabled. The questionnaires used are described in the appendix of this chapter.

Follow-up telephone interviews were conducted with respondents in organizations that had taken part in the study and with a number of teleworkers.

The use of interviews allowed for more open discussion and greater understanding of complex issues than could have been achieved through the use of a questionnaire alone.

In total, 14 organizations agreed to take part in the study both at an individual level (where teleworkers answered questions about teleworking) and at a corporate level (where top management answered questions about teleworking). These organizations were mostly in high technology industries (57%). The researchers carried out follow-up interviews with the managers most closely involved in the management of teleworking in five of these organizations. Overall, 20 teleworkers covering a broad spectrum—self-employed as well as employed by organizations, able-bodied as well as disabled (the sample included: three general managers, three workers in the *telecottage* industry, two software engineers, a consultant programmer, two financial controllers, a training center manager, two IT managers, two secretaries, three self-employed journalists and a self-employed internet marketer)—replied to the questionnaire and six of them were selected on the basis of their experience with teleworking (all had been teleworking for many years) to participate in follow-up interviews. These numbers are small, but they reflect the still small number of organizations that have truly experimented with teleworking in Ireland and the small number of teleworkers who were willing to talk to the researchers about their experience.

ANALYSIS OF FINDINGS

Intermittent home-based teleworking was found to be the most common form of teleworking. Thirty-one percent of respondents teleworked one or two days a week and spent the remainder in the main office. Other respondents were mobile teleworkers (15%), who used teleworking to access data after hours (15%) or they used teleworking in all these different ways (38%). No respondent to the survey was exclusively home-based.

The study showed that teleworkers involved in this work arrangement were highly skilled, key employees (at general management level). These findings are in line with the literature, which states that it is most commonly managerial and professional employees who telework intermittently (Internal Revenue Service, 1996; Murray, 1995; Huws, 1993). The results of this study also are in agreement with the literature with regard to the type of jobs best suited for teleworking—those that involve a high degree of intellectual work (e.g., Stanworth & Stanworth, 1991a).

In most cases, teleworking was implemented on an ad hoc basis and under pressure from employees (38%) as opposed to following a set procedure, as advised in the literature. Only in 23% of cases was teleworking implemented as a result of an organizational directive. Thus, top management commitment existed in 69% of cases, but this initial commitment did not seem very robust when other factors were considered. There was an active champion in only 46% of cases. Furthermore, 69% of respondents said that an awareness program had not been carried out, and there was no official teleworking policy or clearly established procedures in 46% of cases. A pilot run was put in place in only 39% of cases. Overall, only 15% of the companies surveyed had developed a specific program to select teleworkers.

A major problem reported by managers in relation to the implementation of teleworking is the disruption to the communication channels in the conventional office (Kugelmass, 1995) because the telecommuting environment reduces opportunities for this interaction (Fritz, Narasimhan, & Rhee, 1998). In contrast, the findings of this study indicate that only 23% of the respondents to this study felt that teleworking caused a disruption to communication channels in the workplace. This may be because, in the majority of cases, employees in our study spent at least three days working in the office and so remained in touch with the office environment.

SUPPORT, TRAINING AND DEVELOPMENT

It is crucial that technical assistance should be available and that technical problems be resolved quickly (Gray, 1995; Kugelmass, 1995). The most common type of technical assistance available to teleworkers was a helpdesk which 77% of respondents had in place. This did not prevent a number of the teleworkers surveyed to identify a lack of administrative and technical support as an area of concern. Overall, 60% of disabled teleworkers (35% of the respondents indicated that they were disabled) as opposed to 40% of able-bodied teleworkers indicated they were in favor of setting up an official phone-based technical support and mentoring service for teleworkers while 80% of the disabled teleworkers felt there was a need for an intranet-based network to aid communication between teleworkers. This result can be considered in light of the fact that 46% of the organizations did not provide any special training.

TECHNOLOGIES, WORK PRACTICES AND IS SUPPORT

This study found the cost of ISDN lines and the cost of making ISDN calls remain an inhibiting factor to more widespread adoption of teleworking in Ireland. This negatively affects the possible uptake of video conferencing as part of any teleworking arrangement because installing ISDN is viewed as a disincentive and inhibiting factor in video conferencing plans. Video conferencing is used regularly in Irish organizations for transatlantic meetings but is not part of any teleworking plan, and none of the teleworkers surveyed used video conferencing in their work because it was felt to be unnecessary and too costly. In other words, unless the cost of video conferencing equipment is seen as affordable, its adoption may remain minimal. On teleworker who was asked if she felt video conferencing would reduce the sense of alienation experienced by many teleworkers (Nakamura, Ide & Kiyokane, 1996) said it would be of no benefit to her because she felt that it would not be 'real enough' and she preferred to see people face-to-face. This feeling was confirmed by other teleworkers. It is therefore likely that face-to-face meetings will remain an important part of any teleworking arrangement. However, groupware tools were used by 62% of the organizations surveyed and all used e-mail.

RESISTANCE FROM IT STAFF

Contrary to a view expressed in the literature (Warwick, 1996), none of the respondents had experienced resistance from IT staff. This probably is because teleworking is often not seen as a priority by IT management and, therefore, IT staff are not under pressure to provide a high level of support. However, one IS manager mentioned that teleworking was introduced because the technology was available, but he added: "The company does not support working from home as a policy."

RATIONALE FOR THE IMPLEMENTATION OF TELEWORK

According to the literature, the most common reason for adopting teleworking is in response to requests from members of staff (Internal

Revenue Service, 1996). In almost 80% of the organizations surveyed, teleworking was introduced at the request of employees. Many organizations (62%) also realized that teleworking would provide them with access to specialists' skills and facilitate the retention of rare skills.

The most recognized benefit of teleworking was that it offers greater flexibility (92%), but only one-third of the companies in the sample agreed that teleworkers are more productive than their counterparts in the main office.

Reduction overhead costs were identified by 69% of respondents as an organizational benefit, especially through reductions in total office space required. However, *hot-desking* only was used by 31% of the organizations in the sample.

PROFILE OF THE TELEWORKERS

The majority of teleworkers in this study (87%) held a third level qualification (i.e., had obtained a university degree or equivalent) indicating the intellectual nature of their work, mostly of managerial nature. But, most of the people in this survey who were employed in industry currently spend about two days a week at home and the remainder in the office. These people generally had many years' experience in their jobs and were therefore trusted by management.

In this study, both management and teleworkers agreed that the most important characteristics of teleworkers are that they be self-motivated, well organized, trustworthy, and self-disciplined. While 100% of the disabled teleworkers recognized the importance of being trustworthy, management remained reluctant to allow employees to spend a protracted period of time at home without reporting regularly to the main office thereby effectively ruling out teleworking as an option for disabled employees. Despite the views outlined in the literature, only two of the organizations surveyed employed a disabled person in a teleworking capacity.

The majority of able-bodied teleworkers (80%) were employed by an organization, while 60% of disabled teleworkers were self-employed (35% of the respondents indicated that they were disabled). This may indicate that it is more difficult for a disabled person to be employed as a regular employee of a company in a teleworking capacity. This may be because telework generally is restricted to the existing staff of a company.

CONCLUSION

The aim of this research was to explore how teleworking has been implemented in organizations in Ireland and to contrast our findings with existing literature. This study revealed that intermittent, home-based teleworking is the most common form of teleworking. Indeed, none of the respondents to this survey was an exclusively home-based teleworker.

It appears that even where specific policies and procedures exist, teleworking is implemented in an ad hoc manner. The majority of teleworkers claim to have inadequate technical and administrative support (80% of disabled teleworkers and 30% of able-bodied teleworkers). Although 77% of the organizations claimed to provide such support, 46% of teleworkers receive no special training. It seems that top management are reluctant to actively encourage teleworking and, as a result, IT managers do not view teleworking as a priority.

The greatest advantage of teleworking for Irish organizations is the increased flexibility it allows (90% of respondents). Organizations also found that teleworking helped reduce overhead (69%), retain highly skilled employees (62%), and provided access to a wider labor market (62%). There was some evidence, however, that management may view teleworking as a necessity in some cases, but generally be reluctant to encourage its uptake. Thus, it was usually an employee-led initiative (38%), which was implemented more out of necessity than anything else. Teleworkers in this instance were likely to be skilled professionals whose work was of intellectual nature and high value added. Furthermore, 85% of respondents held a university degree or an equivalent qualification.

Generally, the technology required to implement teleworking is present in Irish organizations, but this technology is not being fully exploited. Further improvements and cost reductions need to be made in the Irish telecommunications system (e.g., increasing the bandwidth to allow the introduction of new techniques such as tunneling) that would provide increased data security and allay the concerns expressed by management in this regard.

Based on the testimony of the disabled teleworkers who took part in our the study (35% of all respondents), teleworking can provide employment options for disabled people and should be encouraged. However, since the disabled are not a homogeneous group (Huws, 1996a; British Telecom, 1992), teleworking may not be suitable for all disabled workers and cannot replace the need for conventional workplaces to be made universally accessible. Existing employees who develop chronic illnesses and those with debilitating conditions can benefit from teleworking, as organizations are more likely to grant existing and trusted employees teleworking arrangements. But, the very positive experiences of the respondents to this study suggest that teleworking is a highly suitable form of employment for people who have special needs.

However, some obstacles must be overcome, and Irish managers must be persuaded of the very real benefits inherent in teleworking. Overall, the development of teleworking is dependent on more than just the implementation of required technology. It first and foremost must be concerned with new methods for organizing work and therefore constitutes a new challenge for managers, a challenge that they are unwilling to take on at this point. This is regrettable given that all respondents in this study had only positive experiences to report on the implementation of the concept of teleworking. The highest hurdle in implementing teleworking may very well be the psychological barrier perceived by Irish managers in adapting their methods of supervision and coordination to new organizational forms. Greater awareness of successful experiences with telework through the publication of books such as this one and open debates on the issue between managers and interested workers will undoubtedly help overcome such obstacles.

REFERENCES

Benjamin, N. (1996). The Right Stuff: Teleworking at Lloyds TSB. *Flexible Working*, August, 25-27.

Berry, S. (1996). Teleworking Today. *Computing and Control Engineering Journal*, February, 4-8.

Bertin, I. & O'Neill, G. (1996). *Telefutures*. Ireland: Forbairt and Telecom Eireann report.

Bertin, I. & Denbigh, A. (1998). *The teleworking handbook, new ways to work in the information society*, Kenilworth, UK: TeleCottage Association.

Bonoma, T. (1985). Case Research in Marketing: Opportunities, Problems and a Process. *Journal of Marketing Research*, 22(2), 199-208.

British Telecom (1992). *Disability and teleworking: A report by BT research laboratories*. London, UK: British Telecom.

Brain, D. & Page, A. (1991). *Review of current experiences and prospects for teleworking*. Luxembourg, Luxembourg: Office for Official Publications of the European Communities, 1991.

Burch, S. (1991). *Teleworking: A strategic guide for management*. London: Kogan Page.

Carruthers, S., Humphreys, A. & Sandhu, J. (1992). *Out of the net: Teleworking and people with special needs.* A report for the European Commission's TIDE 166 INDICES Project, Special Needs Research Unit, University of Northumbria, UK.

Chapman, A., Sheehy, N., Heywood, N., Dooley, D. & Collins, S. (1995). The Organizational Implications of Teleworking. *International Review of Industrial and Organizational Psychology*, 10, 100-110.

de Beer, A. & Blanc, G. (1985). *Le travail à distance: Enjeux et perspective, une analyse documentaire.* Paris, France: Association Internationale Futuribles.

Deipser, A. & Stegmann, R. (1996). Teleworkplace for the handicapped, interdisciplinary aspects on computers helping people with special needs. *5th International Conference on Information Systems*, vol. 1, 189-195.

Denzin, N. K. (1978). *The Research Act.* New York: McGraw-Hill.

Ettigoffer, D. (1992). *L'entreprise virtuelle ou les nouveaux modes de travail.* Paris, France: Editions Odile Jacob.

Fowler, A. (1996). How to Benefit from Teleworking. *People Management*, 2, 5-34.

Fritz, M., Narasimhan, S. & Rhee, H. (1998). Communication and Coordination in the Virtual Office. *Journal of Management Information Systems*, 14(4), 7-28.

Goll, S., Lilienthal, T. & Zapp, M. (1998). Teleworking for disabled person, opportunities and risks: An empirical study. *6th European Congress on Research in Rehabilitation*, Berlin, Germany.

Gray, M. (1993). *Teleworking explained.* London, UK: Wiley.

Gray, M. (1995). The Virtual Workplace. *ORMS Today*, August 1995, 22-26.

Hesse, B. (1995). Curb cuts in the virtual community: Telework and persons with disabilities. *Proceedings of the 28th Annual Hawaii International Conference on System Sciences*, Vol. 4, 418-425.

Hewitt, P. (1993). *The Revolution in Work and Family Life.* London, UK: Oram Press.

Huws, U. (1984). *The new home workers: New technology and the changing location of work.* London, UK: Low Pay Unit.

Huws, U. (1993). *Teleworking: Facing up to the future.* Health & Safety Information Bulletin #223, UK: Department of Employment.

Huws, U. (1994). *Social Europe, follow-up to the white paper on teleworking.* Report to European Commission's Employment Task Force, Supplement 3/95, Luxembourg, Luxembourg.

Huws, U. (1996a). *A Manager's Guide to Teleworking*. London, UK: Department of Employment.

Huws, U. (1996b). Teleworking: An Overview of the Research. *Flexible Working*, 1(4), 149.

Huws, U., Korte, W. & Robinson, S. (1990). *Telework, Towards the elusive office*. Chichester, UK: Wiley.

Illingsworth, M. (1994). Virtual Managers. *Information Week*, June, 14-17.

Incomes Data Services Ltd. (IDS) (1994). *Teleworking*. IDS Study 551, April, UK: IDS.

Information Society Ireland. (1996). *Strategy for action*. Report of Ireland's Information Society Steering Committee. Dublin, Ireland: Forfas.

Internal Revenue Service (IRS). (1996). Turn On, Tune In, Churn Out: A Survey of Teleworking. *IRS Employment Trends*, March 1996, 6-15.

Katz, A. (1987). The Management, Control and Evaluation of a Teleworking Project: A Case Study. *Information and Management*, 13, 179-190.

Kling, R. & Sacchi, W. (1982). The Web of Computing: Computer Technology as Social Organization. *Advances in Computers*, 21, 85-97.

Kraut, R.E. (1987). Predicting the use of technology: The case of telework. In R.E. Kraut (Ed.), *Technology and the Transformation of White-Collar Work* (113-134). Hillside, NJ: Lawrence Erlbraum Associates.

Kraut, R.E. (1989). Telecommuting: The Trade-Offs of Home Work. *Journal of Communications*, 39(3), 19-47.

Kraut, R. (1994). *Research recommendations to facilitate distributed work*. National Research Council, Technology and Telecommuting: Issues and Impacts Committee, R. Kraut, Chair, Washington, DC: National Academy Press.

Kugelmass, J. (1995). *Telecommuting: A manager's guide to flexible work arrangements*. Lexington, MA: Lexington Books.

Li, F. (1998). Team-telework and the new geographical flexibility for information workers. In M. Igbaria & M. Tan (Eds.), *The virtual workplace*. Hershey, PA: Idea Group Publishing.

Lindstrom, J., Moberg, A. & Rapp, B. (1997). On the Classification of Telework. *European Journal of Information Systems*, 6, 243-255.

Littlefield, D. (1996). Nationwide Moves Over to Telework. *People Management*, September, 7.

Lynch, T. & Skelton, S. (1996). Teleworking in the Information Society. *Computing and Control Engineering Journal*, February, 33-38.

Markby, D. (1993). Integrating Teleworking Into the Organization. *Computing and Control Engineering Journal*, August, 167-169.

McDonald, S. (1996). Teleworking the New Cottage Industry. *Irish Computer*, 20, November, 9.

Meehan, A. (1997). It's Time to Look at Teleworking in a New Way. *The Sunday Business Post*, 22 June, 22-23.

Murray, B. (1995). The Economic Arguments for Teleworking. *Flexible Working*, November, 31-33.

Nakamura, K., Ide. T., & Kiyokane, Y. (1996). Roles of Multimedia Technology in Telework. *Journal of Organizational Computing and Electronic Commerce*, 6(4), 385-399.

Nilles, J. (1994). *Making telecommuting happen: A guide for telemanagers and telecommuters*. USA: Van Nostrand Reinhold.

Olson, M. H. (1987). Telework: Practical experience and future prospects. In R. E. Kraut (Ed.), *Technology and the transformation of white collar work* (137-152). Hillside, NJ: Erlbraum.

Olson, M. H. (1989). Work at Home for Computer Professionals: Current Attitudes and Future Prospects. *ACM Transactions on Information Systems*, 7(4), 317-338.

Parliamentary Office of Science and Technology (POST) (1995). *Working at a Distance: UK Teleworking and its Implications*. London, UK: Government publications.

Reid, A. (1993). *Teleworking - A Guide to good practice*. UK: Blackwell.

Rogerson, S. & Fairweather, N. B. (1997). Teleworking. *IMIS Journal*, May 1997.

Rothwell, S. (1987). How to Manage from a Distance. *Personnel Management*, 19, 22-26.

Sandhu, J., Humphreys, A. & Carruthers, S. (1992). *Interfacing disabled people to industry-standard computing environments: Report on requirements and functional specifications*. CEC TIDE T166 INDICES, Newcastle Polytechnic, UK.

Sandhu. J. & Richardson, S. (1989). *Concerned technology: Electronic aids for people with special needs*. Research Unit, Newcastle Polytechnic, UK.

Semler, R. (1989). Managing Without Managers. *Harvard Business Review*, Sept-Oct, 76-84.

Simmons, S. (1996). *Flexible Working: A Strategic Guide to Successful Implementation and Operation*. London, Kogan Page.

Stanworth, J. & Stanworth, C. (1991a). *Telework: The human resource implications*. London, UK: Institute of Personnel Management.

Stanworth J. & Stanworth, C. (1991b). *Work 2000*. London, UK: Chapman and Hall.

Taylor, S. J. & Bogdan, R. (1984). *An introduction to qualitative research methods* (2nd Edition). New York: John Wiley & Sons.

Warwick, M. (1996). Home and Away. *Banking Technology*, April, 34-36.

Weiss, T. (1998). *Telework and video-mediated communication: Importance of real-time, interactive communication for workers with disabilities*, Adaptive Technology Resource Centre, University of Toronto, Canada: http://www.utoronto.ca/atrc/ [Dec. 31, 1998].

Wilkes, R., Frolick, M., & Urwiler, R. (1994). Critical issues in developing successful telework programs. *Journal of Systems Management*, July, 30-34.

Yahoo server (1998). Ranking and classification of the top 10 countries in the world [on-line]. Available: http://www.yahoo.com/ [May 12, 1998].

APPENDIX

Structure of Questionnaire for Companies

The questionnaire was distributed mainly by post but also by e-mail in some cases. An effort was made to design a questionnaire that would be quick and easy to complete. Closed questions were used where possible, since they are more convenient for respondents and are easier to analyse. Open-ended questions were used where more information was needed but their use was kept to a minimum. The questionnaire was of the following general format:

Questions 1-3

These questions were asked in order to elicit some background information about the respondent organizations' experience of teleworking.

Questions 4 and 10

These questions related to the issue of employing people with special needs in teleworking arrangements.

Questions 5 - 19, 24 - 26

These questions related to the implementation of teleworking in the organization.

Questions 20 - 23

These questions related to personal characteristics of teleworkers, positive aspects of teleworking for the organization and issues of concern for management.

Structure of the Questionnaire for Teleworkers:

The questionnaire was divided into four sections and was of the following general format:

Section 1: Questions 1 - 7

These questions provided background information such as employment status, level of education, age, employment history and the origin of the individuals involvement in teleworking.

Section 2: Questions 8 - 19

These questions related to the individuals' experience of teleworking and encompassed issues such as the type of work arrangement, hours worked, reason for choosing telework, job type, technologies used, positive and negative aspects of teleworking, important personal characteristics and job satisfaction.

Section 3: Questions 20 - 24

This section covered issues of disability, difficulty in gaining employment, and training received.

Section 4: Questions 25 - 28

This section dealt with possible initiatives that could be undertaken to improve the teleworker's situation. Question 28 gave respondents the opportunity to air any additional issues of importance to them.

Chapter III

TelewerkForum Stimulates Telework in The Netherlands

Kitty de Bruin
TelewerkForum, The Netherlands

ABSTRACT

Implementation of successful teleworking programs requires a cooperation among organizations, employees, governments and labor groups. This paper presents the model used in the Netherlands that can be adopted by any region, state or country to promote telecommuting for the good of society and the employees. The success of any telecommuting program relies on proactive communications, practical advice, and constant promotion of the concept to the public.

INTRODUCTION

The TelewerkForum in the Netherlands is an independent organization set up by the private sector in 1996 and sponsored by the Dutch Ministry of Transport, Public Works and Water Management. As a nonprofit organization, it brings together organizations, unions, and employees whose common interest is to stimulate telework of work in the Netherlands. TelewerkForum's mission is to stimulate the implementation of new forms of working and to contribute to economic and social change by collecting and disseminating information. It supports the transition of the Netherlands into an information society. To implement a successful telework project, an integrated approach is required including office technology innovation, management tools and changing organizational structure.

In the definition of telework used by TelewerkForum, work is characterized as independent of time and place and makes use of information and communication technology. There are several types of telework. In home-based telework or telecommuting, an employee or contractor works at home instead of traveling to an employer's or client's premises. In mobile telework, employees travel from home directly to the client. Telecenters provide access to neighborhood and satellite offices where the employee can work.

The number of teleworkers is projected to rise worldwide due to social developments such as an increase of more highly educated people who are information workers, a growing individualization of people's needs for working, a lack of highly skilled personnel for industry needs, an awareness of environmental issues and limitations, increased traffic congestion and the globalization of company scope. Based on the trends as described, TelewerkForum expects the number of teleworkers in the Netherlands to double in the years ahead.

Telework offers several advantages to companies, individuals and society. A focus point where all relevant information can be found helps companies to implement telework programs and assists employees. TelewerkForum is a small, flexible organization with a board of directors consisting of six members, a managing director and secretarial support. The managing director works approximately three days per week and a secretary works two days per week and spends one day on the web content update (on a contract basis). An advisory committee for the association is comprised of representatives from unions, universities, the Dutch Employers Organization, the Ministry of Economic Affairs and the Ministry of Transport, Public Works and Water Management. The board of directors is chosen among participants and sponsors and meets once a month. There are various working groups comprising participants, experts, social partners, universities and interest groups. The TelewerkForum is set up as a nonprofit organization. This means it does not compete against suppliers, but offers support for the development of telecommuting along with several suppliers and the involvement of the social partners.

The physical office space is shared with one other professional association and uses the support infrastructure (e.g., the reproduction department) of the ICT suppliers to keep overhead costs very low. The public relations and marketing people of the organizations are linked to the work groups via virtual working and teaming.

In December 1995, an initial organizing meeting was held for companies and institutions that might be interested in participating and 15 of them signed to participate for two years. The participants gain from the development of the telecommuting market through sharing knowledge and combining efforts to approach the market. During that meeting the TelewerkForum forecast that the number of teleworkers would increase from 57,000 in 1994 to 441,000 in 2000 in the Netherlands (Lavery & Templeton, 1993). This has turned out to be the case and even more.

The growth projections for the Netherlands that convinced the participants to sign up are shown in Table 1.

The main activity of the forum is to provide information and advice free of charge to companies, government and institutions that are planning to implement telework projects. TelewerkForum collects and classifies information, articles, research papers and experiences and distributes these to target groups. For paying member institutions, TelewerkForum offers the possibility for joint market and product development, for joint research projects and for the exchange of information. Participants can advertise their products and services on the TelewerkForum web site, but they are not mentioned in the printed paper brochure.

The main target group is the employers at the management level. The TelewerkForum uses the following methods to reach them. To provide information for the management-level employer, the TelewerkForum director gives speeches at seminars for employers and target groups such as unions (over 100 in the last two years). Distributing free brochures and fact sheets (20,000 since 1998) also spreads the teleworking concept. Another service is answering questions via telephone and e-mail about the effects of teleworking

Table 1: Projections for telework growth and related business in the Netherlands

	1994	2000
Number of teleworkers in the Netherlands	57,000	441,000
Hardware products	$ 21 million	$164 million
Software products	$ 20 million	$361 million
Telecommunication products	$ 11 million	$126 million
Data communication traffic	$3.5 million	$ 17 million

(Lavery & Templeton, 1993)

for employers. The forum also provides information about social, fiscal, labor security and technical aspects to employers by sending articles on the specific subject to anyone requesting the information. The director publishes six newsletters per year in the magazine *Telewerken* (the Dutch word for telecommute). This magazine has a circulation of 4,000 copies, and is distributed to managers in Dutch organizations. Another item has been the organizing of participation in projects such as the traffic reduction feasibility study. The TelewerkForum did the fund-raising and participated in the steering committee to find the right partner to conduct the investigation, and presented the results to a mixed audience—press, politicians, unions and companies.

Most companies are looking for guidance on, and good examples of, practical telework program implementation. To recognize success and promote the concept, the TelewerkForum presents annual awards to the projects that best served as examples for other organizations. The Telework Aspect Prize is given for the best product or service that stimulates telework. The selection jury consists of high-level representatives of the Dutch Chamber of Commerce, small and medium enterprises, the Dutch Employers Organization, two trade unions, the Ministry of Transport, Public Works and Water Management, the Ministry of Social Affairs and Employment and the Ministry of Economic Affairs. Last year more than 50% of the projects involved small and medium enterprises, compared to earlier years when the projects of large organizations dominated the awards. In 2000, the Telework Awards will be presented by one of the government ministers at a seminar.

The web site is regularly updated with the latest news and information concerning seminars held on the various aspects of telework. The track record of hits on Frequently Asked Questions (FAQs) is published in the newsletter and on the web site. The web site provides a Quick Scan for employers and employees to determine feasibility for telework. The discussion forums on the web site are monitored, and views and answers are given. From March 1999 to March 2000, the average hit rate was 1500 hits per day, with a peak in the last month of more than 5,000 hits a day, following a large radio and television publicity campaign about telecommuting by one of the participants. In 1999, work began on building a knowledge database of more than 300 publications, classified and retrievable per subject. This database will have links to the universities that have been asked to classify current and future research projects per subject. This will enable identification of subjects for further research to be done in the Netherlands.

The director of the TelewerkForum participates on committees about various subjects such as the building of modern accommodation suitable for telework, new accommodation complexes, mobility and the combination of work and care. A lot of cooperating is done with traffic coordination centers by holding combined workshops and seminars. Support to and from these centers will be extended in the future. A good relationship has been built up with institutions for the reintegration of disabled people.

A press contact database has been created and many interviews are given for radio, TV and magazines. In 1999, there were more than 100 articles in trade journals, life-style magazines and the newsletters of the unions and companies. The TelewerkForum gives advice about telework to the editors of brochures and manuals. The unions, who were not in favor of telework several years ago, have changed their views. A representative of the CNV, the Christian Labor Union, is a member of the TelewerkForum public relations workgroup.

After intensive lobbying by TelewerkForum, a new fiscal rule was introduced in the Netherlands in January 1998. Companies are now allowed to give employees an annual tax-free sum of approximately $400 to buy furniture for home-based telework or a lump sum of $2,000 to cover the five years ahead. In the original proposal sent to the ministries in 1995, TelewerkForum called for tax-free compensation to a maximum of $2000 per year. This amount had been calculated by the council of employers of one of the large bank organizations. In the opinion of the Forum members, the $400 result is too low and only appropriate for teleworkers who work one day at home, not for mobile workers who work from home and visit clients every day. There is no specific labor law statute concerning telework, but the 1994 Working Environment Act and the Decree on Home Employment are officially applicable to work done in the home environment (Pennings, van Rijs, Jacobs, & de Vries, 1996). Most employers make individual contracts concerning telework with employees, and telework has been mentioned in one of the collective labor agreements for the first time in 2000.

TelewerkForum was used as a model to set up similar telework associations in France and Germany. The TelewerkForum director was involved in the action plan to stimulate telework in the European Telework Development Program. Romania has shown an interest in TelewerkForum and presentations have been made in Finland, Germany, France and Belgium. The European Telework Development Program (1995-1999, funded by the European Commission, Directorate—General Information Society) stimulated the exchange of information between existing telework associations and the founding of associations in countries without telework information points. The national coordinators of this program were mostly the members of the telework associations, and they provided part of the input for the Status Report on European Telework (Johnston & Botterman, 1999). This report and information concerning the European Telework Associations can be found on the Internet at www.eto.org.uk.

TelewerkForum raised funds to conduct a feasibility study into the possibility of telework for 10,000 people living in the Amsterdam-Utrecht-The Hague triangle as a means of reducing traffic congestion. The study was published and presented to press, politicians and companies in June 1999. The research indicates that it is realistic to aim at more flexible working arrangements within a few years for 10,000 employees in the geographical triangle Amsterdam-Utrecht-The Hague. Figures from the past have been unable to bring out that this will reduce traffic jams in the triangle within that period, but it certainly supports the current social trend that is leading to many more people avoiding traffic jams related to their work. The methodology used included desk research, in-depth interviews and an electronic questionnaire on the Internet.

The study examined four main aspects of telework: realistic size of teleworker body, possible traffic jams reduction, business motivations, and technical feasibility. The study's conclusion is that 25,000 potential teleworkers is rather ambitious, but removing a total of 10,000 teleworkers from traffic jams in the geographical triangle is feasible. The figure is based on data from a morning rush hour in October when 184,912 people were caught in traffic jams in the triangle, 85% of them because they were commuting or traveling on business. Thirty percent of this group, i.e. 47,175 people, would be suitable candidates for teleworking. A percentage of between 10% and 20% of this figure is much more realistic than the ambitious 50% originally envisaged. This still means that a significant number of people would make the move from asphalt to the electronic highway. The feasibility of the move from physical to virtual relocation is further strengthened because it links up with a number of important developments within organizations in the Netherlands.

A rough indication of the number of people caught in traffic jams during the morning rush hour in the geographical triangle is 185,000. In theory, it has been proven that a reduction of 30,000-35,000 drivers in the triangle during the morning rush hour would have a noticeable effect on the traffic jams. This effect would be identical to the difference between the traffic jams formed in the July morning rush hour and those of the

October morning rush hour. In addition, teleworking results in a different mobility pattern in practice, with teleworkers avoiding traffic jams. Teleworking for a few days a week results in a reduction of 6-15% in the number of kilometers driven in the rush hour.

Businesses have an eye to their own interests where teleworking is concerned and are a fruitful breeding ground for realizing the traffic reduction plan. Their interests include keeping and attracting quality personnel, addressing the problem of accessibility and mobility, and increasing attention to the balance between work and private life. Teleworking, within the framework of new types of work, is encouraged from the point of view of increasing flexibility at work and in the organization.

The results of the survey indicate wide support for participation in the traffic reduction plan. No fewer than 96% of the 449 respondents would like to telework, and 62% see no insurmountable obstacles in the nature of their work activities. In concrete terms, 70% of the respondents consider it possible that their organizations will participate in the traffic reduction plan. These data support indications that teleworking in practice has undergone such fundamental changes over the last six years that they are more accurately characterized as a metamorphosis.

Concerning the question of whether the traffic reduction plan is feasible technologically, the general conclusion is that the technological feasibility of teleworking is in the hands of the organizations themselves. Implementation of teleworking is usually seen as a different approach to technological elements in the existing IT infrastructure of an organization. Technological developments are developing rapidly in a way that increasingly facilitates accessible and large-scale use of IT.

The four year TelewerkForum plan covers the three phases of the traffic reduction plan: the incubation phase, the realization phase, and the consolidation phase. The results of the in-depth interviews indicate that a throughput time of two years is not feasible, given the fact that each participating organization needs a tailor-made approach. The Forum's aim is to find companies where 20% of the employees can participate in teleworking.

The goal of TelewerkForum over the next two years is to identify five large companies or organizations with 1000 new teleworkers each, five large organizations with 500 new teleworkers each, 15 medium-size organizations with 100 new teleworkers each and 20 small- to medium-size enterprises with 50 new teleworkers each. A follow-up action plan is currently under consideration by TelewerkForum and the Ministry of Transport, Public Works and Water Management.

The TelewerkForum focuses on providing information to the employer and the employees. However, many questions come from unemployed people or persons who are not allowed by their employer to telework. The most frequently asked question on the web site is, "Where can I find telework?" The TelewerkForum, together with the Ministry of Social Affairs and Employment, is conducting a feasibility study into the possibility of setting up an employment agency for teleworkers. The aim of this agency would be to match the skills of unemployed people willing to telework with the market demand. The idea is to route the unemployed looking for jobs to another web site, and to keep the original telework site exclusively for companies that want to implement telework. More than 640 people have been interviewed and the response of more than 200 shows that there is a large potential of reasonably educated people, mostly with more than 10-15 years of work experience, who are willing to work for 10-20 hours per week in a telework mode.

TelewerkForum has been able to attract sponsorship from companies and governmental ministries that would benefit most from the development of telework. Since the start in 1996, TelewerkForum now has more than 20 organizations that not only provide financial support, but also participate in working groups. The organizations include IT suppliers, consulting firms, cable providers, office furniture suppliers and a research institute. The annual budget is low (in comparison to the organization's goals) and funded by the participants. Additional funding from nonparticipants is raised for the special projects (e.g., the Telework Awards or feasibility studies). Currently the budget is US $150,000, based on the participants paying US $5,000 each and the sponsors between $15,000 and $25,000. Negotiations are under way with the Ministry of Transport and Water Management to discuss expansion of budget and manpower. In these negotiations, outcomes such as the traffic reduction plan and awareness campaigns to bring telework to the attention of employers are being discussed.

The number of employed teleworkers in the Netherlands has increased from 57,000 in 1994 to between 600,000 and a million in 1999 (Johnston & Botterman, 1999). This is more than 14.3% of the total workforce. The Netherlands has the highest percentage of teleworkers in Europe. This is because of the culture, where the distance between the boss and employee is low, the high level of adoption of technology, and a high proportion of jobs in services (the highest in Europe). The Netherlands was the first country to liberalize the telecom industry and now has more telecom providers than Japan. Living standards are also high and the country has progressive labor market policies, bringing a strongly analytical and innovative approach to addressing labor market issues. An example is the law of flex and certainty. Employees now have the right to work 20% less than a full week. A successful economy has led to a shortage of skills, providing motivation for companies to embrace new methods of work, and several companies are now using telework as a perquisite to attract new personnel. One major sponsor has broadcast advertisement campaigns on radio and TV to reinforce the concept of telework. Many companies interested in implementing telework have requested information and have copied parts of the web site, such as Quick Scan, for their own .

ENDNOTE

Additional information on TelewerkForum is available at http://www.telewerkforum.nl.

REFERENCES

Johnston, P. & Botterman, M. (1999). *Status Report on European Telework: New Methods of Work*, European Commission Directorate-General Information Society.

Lavery, M. & Templeton, A. (1993). *Flexible Working with Information Technology*. London: OVUM Ltd.

Pennings, F. J. L., van Rijs, A. D. M., Jacobs, A. T. J. M., & de Vries, H. H. (1996). *Telework in the Netherlands*. Amsterdam: Hugo Sinzheimer Institute.

Chapter IV

Public Communications Infrastructure Support for Telecommuting

Joseph R. Bumblis
U S WEST, USA

ABSTRACT

As the number of telecommuters grows, so does the data traffic on the Public Switched Telephone Network (PSTN) and its adjunct infrastructures. This chapter analyzes the impact of increased telecommuter-networking traffic on the PSTN and the adjunct infrastructures such as the Internet. This analysis will give readers the tools to understand the public network infrastructures and to gain insight into the networking requirements of virtual offices/work groups. By understanding the interconnection issues, decision-makers will be better able to analyze business needs related to telecommuter remote offices. In addition to understanding telecommunications technology, corporate management also will gain insight into the language of the service provider. This will help management and key decision-makers evaluate benefits of a telecommuting workforce, and negotiate with service providers for the necessary technology to support of their telecommuters.

CORPORATE SERVICES
AND INFRASTRUCTURE

Most corporations have internal network infrastructures to support corporate R&D, finance, accounts payable/receivable, and general business activities. The most common network components generally include Ethernet as the network medium/transport, TCP/IP as the internetworking protocol, centralized and distributed servers for data and application repositories, and desktop PCs with local applications for word processing, spreadsheet-based calculations, and presentation compilation. Telecommuters require similar services from corporate infrastructures while working from their homes. The following sections briefly explain the typical corporate network components as an introduction to addressing telecommuter requirements.

Ethernet—Ethernet is accepted throughout the computing industry as the local area networking technology of choice. The main attributes of Ethernet include:

- Standardized by the IEEE and ISO as IEEE802.3 and ISO8802.3 respectively,
- Capable of up to 10 Mb/s burst transmission data rate,
- Capable of using coaxial cable, twisted-pair copper cable, fiber optics, and wireless transmission mediums, and
- Available for any platform at costs around $70 per connection.[1]

Since the original IEEE802.3 standard published in 1986, the IEEE has increased the scope of Ethernet to include 100 Mb/s and 1000 Mb/s (1 Gb/s). This was done to support high-speed connectivity to the desktop while maintaining the consumer's investment in cabling/wire plant and application software and drivers.

Internet Protocol (IP)—IP was initially developed as part of the original DARPA network around the early 1970s. It since has been enhanced and currently is the worldwide solution of the Internet. IP is a connectionless (meaning the client computer does not have to establish a session with the server) delivery protocol that performs addressing, routing and control functions for transmitting and receiving information over a network. As packets (variable length segments of information) are received by the router, IP addressing information is used to determine the best "next hop" the packet should take en route to its final destination. As a result, IP does not control data path usage. If a network device or line becomes unavailable, IP provides the mechanism needed to route client or server information around the affected area.

Transmission Control Protocol (TCP)—As with IP, TCP also was developed originally as part of DARPA initiative. TCP guarantees end-to-end delivery of packets and provides a reliable, connection-oriented transport layer link between two network entities (TCP establishes a session between client and server or between server and client similar in function to making a telephone call). Using a two-way handshaking scheme, TCP provides the mechanism for establishing, maintaining, and terminating logical connections between hosts using IP as their network protocol.

The telecommuter and management may want to investigate which TCP/IP extensions are implemented on the corporate routers and servers. Based on the information received from the Network Administrator, the telecommuter's home router and/or computer should be configured for the same options. Note that these options may require a faster service than a dial-up service via the PSTN. Better performance can be realized with higher bandwidth services like T-1, Frame Relay, ISDN, ADSL, or Cable Modems. These technologies will be discussed in more detail later in the chapter.

User Datagram Protocol (UDP)—UDP provides a connectionless (i.e., no telephone call is required; functionally more like a pager) datagram delivery service (transmission of information) between IP host applications. This protocol is used for transaction-oriented utilities such as Simple Network Management Protocol (SNMP) and Trivial File Transfer Protocol (TFTP). Like TCP, UDP encapsulates information into a datagram structure, works with IP to transport the message to a destination, and provides protocol ports to distinguish between software applications executing on a single host. Unlike TCP, UDP avoids the overhead of reliable data transfer mechanism by not protecting against datagram loss or duplication. Thus, UDP is less secure and offers a less reliable network connection than TCP.

BACKGROUND

Public Switched Telephone Network (PSTN)

Before decision-makers can accurately assess a telecommuting work force, they must have a basic, fundamental understanding of the supporting public infrastructure. First, it is essential that basic concepts be reviewed. Figure 1 depicts the interconnect topology of a typical local telephone network, often called the Public Switched Telephone Network (PSTN).

In Figure 1 a caller firsts connects to the End Office (circle) by dialing a

Figure 1: Public switched telephone network (PSTN)

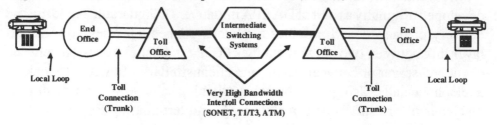

telephone number. End Offices are usually in the Local Access Telephone Area (LATA). If the call is local (within the LATA), the End Office switching equipment simply rings the called number to establish the call connection. If the number is outside the LATA (long distance), then the End Office connects to the Toll Office (triangle) and eventually is connected through the Intermediate Switching System (hexagon) to the called End Office. Toll Offices and Intermediate Switching Systems comprise an Inter-LATA System.[2]

More detail on telephone switching systems can be found in Tanenbaum (1988), Doll (1978) and McConnell (1999). For now, it is important simply to understand that there are many electronic paths (a.k.a. Routes) a call can take to reach a destination. This will become extremely important when we begin exploring how telecommuters would use this system to call an Internet Service Provider (ISP) to establish an e-mail or Internet connection to departmental services physically located at the corporate site.

The PSTN was designed and deployed (with many enhancements over its lifetime) to provide voice services to telephone company customers. Over the past several years, the US population not only has become home computer literate, it has been bombarded with new services like ISP, cellular telephones, pagers, and wireless data services like CDPD and PCS/GSM. These changes have forced telephone companies to segment areas into smaller groups to accommodate the increase in telephone numbers. As a result, all major cities have new area codes that enable telephone service providers to add telephone numbers to service their customer's needs. These needs include cellular telephone numbers, two or more telephone lines in private homes, and pager telephone numbers, all of which have stressed the current PSTN.

An interesting side note is that telephone numbers are very similar in organization and function to Internet Protocol (IP) addresses. Although the Internet has not yet been explained, a brief explanation at this point would be valuable. An IP address is defined as two parts: a network number and a host number. Thus, if a personal computer's attributes were examined, a number could be found that looks like: 139.23.5.111 (some computer systems display all 12 numbers: 139.023.005.111). This is a Class B address, which means the number 139.023 identifies a unique network (the network number) and the number 005.111 (the host number) identifies a unique host (or node) on the 139.023 network.

The corollary is a telephone number. All U.S. telephone numbers comprise two parts: an area code and a unique seven-digit number. The seven-digit number identifies a customer (or node on the telephone network) within the unique three-digit area code. Thus the telephone number 652-555-1234 has a 652 area code (which corresponds to the IP network number) and a customer telephone number of 555-1234 within the 652 area code. The 555-1234 telephone number corresponds to the host number in IP.

An increasing population of telecommuters will further stress the PSTN, regardless of new area codes and digital switches, because the PSTN was designed for short duration voice calls. Telecommuters probably will place calls to an ISP or to a remote server at the supporting corporation, and remain connected for several hours per call. In addition to longer call duration, new modem technologies permit the exchange of digital information at rates exceeding 50,000 bits per second. The PSTN was designed to transmit and receive information at 3,200 cycles per second, or about 1,600 bits per second[3] digitally. It is easy to see that current computer systems are sending and receiving information at almost 400 times the designed limit. Although computer systems are transmitting and receiving data at higher rates than the PSTN was designed to pass, the telephone companies are not affected by this increase. To the telephone switching systems, each connection is only using 3,200 cycles per second of bandwidth (Noll, 1999). A much larger impact is the duration of the calls. A typical voice telephone call is measured in minutes, whereas an average computer or data call usually is measured in hours.

Before evaluating the public systems and whether a corporation can support a large telecommuter population, it is necessary to understand how adjunct infrastructures are used to support the telecommuter. Adjunct infrastructures allow connectivity to services outside the normal reach of the PSTN. Such infrastructures include the Internet, SONET/ATM, wireless, and all fiber/optical networks (AON). Although it is beyond the scope of this chapter to dwell on these technologies in detail, their key attributes will be addressed to help decision-makers evaluate telecommuter feasibility and understand the service provider's language for possible corporate infrastructure negotiations.

The Internet

The capitalization of "Internet" is common in literature to distinguish between corporate intranetworks and the network developed by DARPA:[4] the Internet. In its infancy, the Internet comprised many ad hoc systems consisting of dial-up lines and a few dedicated trunks. Since the early 1970s, the Internet has migrated to high speed backbone[5] technologies capable of tens of gigabits per second (1×10^{10} bits per second or 10 Gb/s) transmission speeds using fiber optic technology. Such technologies include Asynchronous Transfer Mode (ATM), Synchronous Optical Network (SONET), and All-Optical Networks (AON) as described in Ryan (1998) and Pan (1999).

Figure 2 depicts the Internet-1999. Note that PSTN still exists as the common path between ISP clients and the Internet.

The large oval at the top of Figure 2 depicts what is commonly called cyberspace. Its borders encompass the Internet, ISP facilities, PSTN-LATA, and PSTN-interLATA service providers.

The drawing in Figure 2 depicts several interconnect technologies. PSTN Trunk technology such as T-1 (1.544 million bits per second, a.k.a. 1.544 Mb/

Figure 2: The Internet-1999 (adapted from Goldman, 1998)

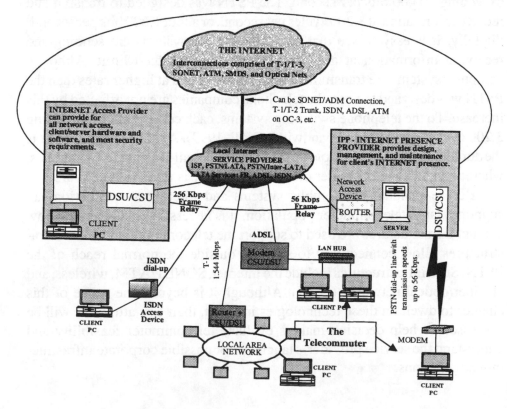

s) and T-3 (~45 million bits per second, a.k.a. ~45Mb/s) still are widely deployed. Individual users (customers) can lease a T-service from the LATA telephone company and connectivity to an inter-LATA service also can be obtained. The most common telecommuter configuration (at least for the next few years) will probably be as depicted in Figure 2. Telecommuters will use the PSTN to dial-in to an ISP for e-mail, Web, and FTP access back at the corporate office. A very good explanation of the Internet, its services and technologies can be found in Goldman (1998). However, wise decision-makers will want to evaluate other possibilities for connecting telecommuters to the Internet. The following sections describe those options. They include: ISDN (dedicated and dial-up), ADSL, cable modems, frame relay, and T-services. An emerging technology, virtual private networks (VPN), currently is being discussed by researchers and Enterprise Network Planners as the next generation enterprise network. Although a detailed discussion of VPNs is beyond the scope of this chapter, a short introduction is provided at the end on this section.

Integrated Services Digital Network (ISDN)

ISDN (Integrated Services Digital Network) has become a ubiquitous technology and is classified as a to-the-home service. As such, telecommuters can obtain ISDN service from their local telephone company for an installation fee and subsequent monthly charges. The monthly fee is a bit higher than a second telephone line, but an ISDN connection can supply connectivity services up to 144,000 bits per second (144 Kb/s). This service requires an ISDN-capable device between the user PC and the ISDN line, which usually is a router with an Ethernet connection and an ISDN connection comprising a configuration as shown in Figure 3. Note the introduction of the acronym "LEC" in Figure 3. Throughout this chapter the local telephone company/ service provider was described as the LATA; and long-distance service providers were referred to as inter-LATA. Within the industry and literature, a service provider within a local area is often called a Local Exchange Carrier (LEC). The inter-LATA (or long distance telephone service provider) is called an IXC (inter-exchange carrier) service provider. Examples of LECs include Bell South, U S WEST, Southwestern Bell, and GTE. Examples of IXCs include AT&T, Sprint, MCI, and GTE.

There are two types of ISDN service: ISDN dial-up service and dedicated ISDN service. ISDN dial-up service is a viable option for connecting to the Internet at speeds up to 128 Kb/s; or about four times faster on average than standard analog modems. If the telecommuters are not constantly using the

Figure 3: A private home ISDN system

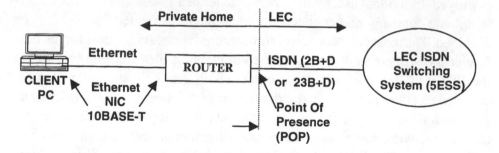

Internet, this shared connection should be adequate for up to 10 or so connections depending on each connection's bandwidth demands. This would permit telecommuters to have several devices connected to the Ethernet side of the router such as printers or file services. When the primary workstation (PC) is not sending/receiving data, the other devices could use the ISDN connection. ISDN service now is fairly inexpensive with most telephone companies charging between $30 to $50 per month for basic service (McConnell, 1997a). Initial installation fees can be as high as several hundred dollars, but this fee differs with each LEC.

Dedicated ISDN Service is necessary if a 7x24 connection to the Internet is required; however, other connection options like Frame Relay or a T-1 line also could provide this service. A Dedicated ISDN connection would require the telecommuter to keep the ISDN connection active 24 hours a day. Most telephone companies charge per-minute tolls for local ISDN calls, usually less than 1 cent per minute; however, this will add up with 24-hour-a-day connections. A typical monthly phone bill for this type of service is typically between $200 to $800 depending on the LEC/phone company's rate structure. Not an inexpensive service, but it is a reduced expenditure when compared to Frame Relay or T-1 (McConnell, 1997a).

In Figure 3, the terms "2B+D" and "23B+D" are introduced. Within the CCITT[6] Standards for ISDN, the signaling rates and connection functionality are defined. The "B" channel (a.k.a. Bearer Channel) is described as a data channel capable of supporting a data rate of 64,000 bits per second. The "D" channel (a.k.a. Delta Channel) is defined as a signaling channel used for call setup, teardown, functionality, and data flow. It is a 16,000 bits-per-second signaling rate channel. Thus with a router between the Client and the LEC ISDN POP (point of presence), the Client can realize a data rate of "2B+D," which equals 2x64,000 + 16,000 or 144,000 bits per second; or "23B+D," which equals 23x64,000 + 16,000 or 1,488,000 bits per second.[7] As expected,

Figure 4: ADSL configuration with POTS (plain old telephone service) connection

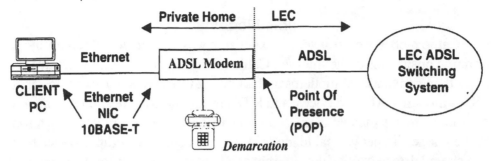

the 23B+D service is quite a bit more expensive than the 2B+D service. For current applications, the 2B+D service should be more than adequate to accommodate the needs of telecommuters.

Asymmetrical Data Subscriber Line (ADSL)

Another Internet adjunct technology making inroads into the home market is ADSL. The xDSL technologies (Goldman, 1998) use digital signal processor (DSP) technology to send data across most home telephone lines at rates approaching 52,000,000 bits per second (52 Mb/s). ADSL technology supports data rates from the service provider to the home at around 6 Mb/s. From the home to the service provider, ADSL supports 640,000 bits per second (640 Kb/s). Both of these rates are supported when the cable length between ADSL modems does not exceed 12,000 feet. Slower data rates are available at longer cable lengths. The typical configuration is similar to the ISDN configuration.

A noteworthy item in the ADSL configuration above is the option to use the homeowner's telephone service (often called POTS for plain old telephone service) in conjunction with the ADSL service. Since the ADSL modem incorporates DSP technology, a voice channel and a data channel can coexist on the same pair of copper cables. For example, telecommuters could have telephone service and a 760 Kb/s data service on the existing home wiring. ADSL is an adaptive service, meaning it adapts to the conditions on the telephone line. If telecommuters have a clear (low noise) phone line and a short cable run to the LEC central office, the ADSL modem will provide speeds comparable to T1 or faster. If the telecommuter's home has a noisy line or is far from an LEC central office, ADSL will run at a slower speed, but still a lot faster than ISDN or analog dial-up service (McConnell, 1997). The basic issue remains that the "LEC ADSL Switching System" infrastructure is far from complete, let alone ubiquitous. Trials currently are in progress with a few live services available.

Cable Modems

Cable modems offer a very inexpensive way to connect telecommuters or home businesses to the Internet. Cable modems operate somewhat differently than conventional modems and ADSL devices. Whereas conventional PSTN modems, ADSL devices, and ISDN terminators are designed to attach or connect to a dedicated circuit that runs from the telecommuters' location to the phone company, a cable modem uses the cable TV company's system as a shared data network (like ethernet or a voice intercom system). All of the devices connected to the network can talk to and listen to each other.

Cable TV systems are designed to deliver a lot of information (moving pictures and audio) from the head end (central distribution point) to users (television sets). These networks are capable of carrying large amounts of computer data in the downstream direction (cable ISP to telecommuter). They usually do not carry as much information in the upstream direction (telecommuter to cable ISP). This is the same situation as with ADSL technology. Telecommuters can receive large volumes of data quickly, but can only send small amounts of data at a somewhat reduced transmission rate.

One problem with cable modems is variability in speed. If many users are using the network simultaneously, the physical and perceived connection speed will decrease. It is basically impossible to precisely predict connection speeds. On average, cable modems deliver between 300 Kb/s to 1.5 Mb/s (McConnell, 1997b). Thus the main advantages are low cost (slightly more than dial up service) and continuous connection. The primary disadvantages are variable service quality and limited availability if the cable hosts a large number of users.

Other Telecommuter Home Connectivity Options

Frame Relay Service generally is regarded as the next step up from ISDN. It runs between 56 Kb/s to 1544 Kb/s (T-1 speed) and usually is available for $1,000 to $3,000 per month, depending on the connection speed, local telephone company and ISP. Frame relay rates can vary widely, so it is a good idea to investigate all possible options (McConnell, 1997c). For example, if both frame relay and ISDN are part of the LEC service package, then the choice made may be for an optimal telecommuter solution as a function of speed, availability, cost, general use (required data traffic), and desired quality of service (QoS).

Quality of service (QoS) has not been discussed in any detail up to this point. Generally, a dedicated connection like T-1/T-3 and Dial-Up over the PSTN have a QoS measured in live, uninterrupted connect time. In other words, is the connection available when the telecommuter wants or needs it? If yes, the service is said to have 100% availability.

Other measures of QoS may influence your choice of network services. Examples of QoS measures include Bit Error Rate (BER), Packet Error Rate (PER), Latency, and Delivery Guarantees. BER is the measure of the probability that a bit of information will arrive at the destination node in error. A common BER is $1x10^{-9}$ which translates to an expected BER of 1 in 1,000,000,000 (one billion). This means that the network service should experience no more than 1 bit out of 1 billion in error.

PER is a measure of the network BER and the switching systems comprising the path between originator and destination (i.e., telecommuter to corporate office servers.) The more intelligence built into the network, the higher the service quality. This generally is true because telecommuters communicate to network services and ultimately corporate services through communications protocols like TCP/IP. These protocols 'packetize' information into variable length packets that are sent across the network to a destination point. If the LEC/ISP offers a high QoS with a low PER, then the network infrastructure has equipment to help guarantee packet delivery. This is accomplished by error checking the packet as it traverses the network from source to destination. This feature is transparent to telecommuters, but may be very important when you select a network service.

T-1/DS1 and *T-3/DS3 Services* were mentioned during the PSTN discussion above (see Figure 2). If the requirement is for a fast (1.544 Mb/s) connection to the Internet, T-1 is a widely used high-speed Internet connection. A DS3 connection offers close to 45 Mb/s. A T1 connection to your ISP will cost (at the time of this writing) $1,500 to $3,000 per month, again

Figure 5: T-1, T-3, or frame relay connectivity

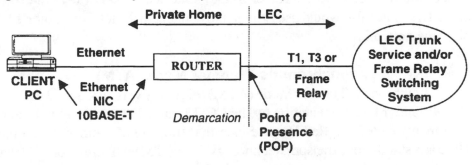

depending on the local telephony company and your ISP (McConnell, 1997c). A DS3 connection could exceed $10,000 per month depending on distance, LEC service pricing and the ISP vendor.

Other Internet Adjunct Infrastructures

In addition to dial-up services over the PSTN, dedicated T-1/T-3 trunks, ISDN connectivity, ADSL connectivity, and cable modem options, there remain numerous technologies that comprise adjunct networks to the Internet. Since these technologies may or may not be available to the individual homeowner, we will not dwell on them. The decision-maker need only be familiar with their attributes and functionality to understand what impact they may have on deciding whether or not to support a telecommuter work force. These technologies include SONET, ATM, FTTC, HFC, AON, and VPN. The next few paragraphs will outline the major attributes of each technology.

Synchronous Optical NETwork (SONET)

Communication between various localized networks is costly due to differences in digital signal hierarchies, encoding techniques and multiplexing strategies. For example, the T-1 (a.k.a. DS1) signals consist of 24 voice signals and one framing bit per frame with a signaling rate of 1.544 Mb/s (Mahal, 1999).

SONET was initially defined by the CCITT as a high-speed trunk for delivery of many different types of telephone services. Since the trunk lines (T-1 and T-3) were already very common within the infrastructure, SONET was designed to carry any and all digital services. SONET is a synchronous "pipe" divided into speeds of 51.84 Mb/s (OC-1).

None of the SONET equipment would be physically located on the telecommuter's property. If the telecommuter required a direct connection into SONET, it more than likely would be in the form of a T1 connection to the home with a direct connection to a SONET connection point. The telecommuter would need a router in the home with a DS1 (T-1) interface card. The router also would use Ethernet to connect to the telecommuter's PC or home network.

Asynchronous Transfer Mode (ATM)

Asynchronous Transfer Mode (ATM) originally was introduced with the ISDN standards documentation by the CCITT (now ITU) in 1984 in a set of documents called the Red Book. Since that time ATM reemerged around 1990 as a stand-alone transport technology. The ATM technology of 1999 has received an enormous amount of publicity as a high-speed, multimedia transport technology (Bumblis, 1997). Although ATM has solved many IXC problems dealing with scalability, ATM remains a second or even third choice for campus and desktop networking solutions behind 10/100 Ethernet and FDDI. This is due in part to the cost of the technology as well as the need to create new application software. By using 10/100 Ethernet and FDDI, many corporations have preserved their investment in application software.

ATM technology reduces traditional Local Area Network (LAN), Metropolitan Area Network (MAN), Wide Area Network (WAN) traffic, and digitized audio and video (multimedia traffic) into a collection of small frames called *cells*. This approach was believed by network architects to hold the ultimate solution to increased network performance, scalability, traffic volume management, and QoS demands. As implementations emerged, this belief slowly faded to the realization that ATM required many modifications before it would be accepted as a LAN solution.

The ATM architecture initially was defined by the ITU (International Telecommunications Union) and described in the Q93B standard. Unlike the variable length packet sizes of Ethernet, Frame Relay, and ISDN, ATM requires all data to be packed in multiple 53-byte cells. This fixed size significantly contributes to the scalability of ATM when used as a WAN solution.

The primary migration plan of the ATM Forum (a vendor consortium) was to allow network planners to maintain their current investment in software (and in some cases hardware) by making ATM "look-like" Ethernet. This was required to support legacy TCP/IP stacks and associated applications, while allowing the enterprise network to begin migrating to an ATM backbone. Telecommuters would be concerned with ATM only if this were a technology used as a Home Area Network (HAN), see "Future Trends" below. Within network infrastructures, ATM Technology is well hidden from telecommuters in the form of the IXC infrastructure as depicted in Figure 2.

Fiber to the Curb (FTTC)

FTTC is an infrastructure technology. As bandwidth demands increase, the infrastructure must be upgraded from the copper cable (often called the Wiring Plant by LECs and IXCs) used today to the higher bandwidth capable with fiber optic technology. For example, this chapter has outlined two high-speed technologies that operate predominately on copper cable: xDSL and cable modems. Based on DSP (digital signaling processor) technology, cable quality, and distance, the best data rate available appears to be around 52 Mb/s (HSDL). Compare this to fiber optic technology, which can span distances exceeding that of copper between users and offer data rates in excess of 10Gb/s. In case the reader needs a refresher on nomenclature, 10 Gb/s means 1×10^{10} bits per second or 10 billion bits in one second.

The goal is to have fiber optic technology at the curb of each homeowner. Thus the name, "Fiber To The Curb." Since this requires a large capital investment by the LECs, most FTTC installations are appearing in new developments only. It is unclear at the time of this writing when, and if, FTTC will be retrofitted into existing communities.

Hybrid Fiber Coax (HFC)

HFC is an extension of the cable TV infrastructure combining both coaxial cable technology and fiber optic technology. The primary goal of most vendors is to use the cable TV-cable already in many homes throughout the United States, and connect a fiber optic infrastructure to provide such services as telephone, video, movies, and internet connectivity.

To many telecommuters, this technology may provide a very reliable connectivity option to support LAN-like performance from the home. Although not widely deployed at the time of this writing, HFC is a technology of great interest to vendors wishing to realize the opportunities of the Telecommunications Act of 1996.

All Optical Networks (AON)

All Optical Networks currently are in trial situations (Ryan, 1998; Pan, 1999). Although many IXC infrastructures incorporate fiber optic technology, none currently offer All Optical Network Services (AONS). The reader is urged to investigate these trials. Should AONs become ubiquitous, telecommuters will realize great strides toward connectivity and "in office" network performance. A very good reference can be found at the MIT Lincoln Laboratories web page: www.ll.mit.edu/aon.

Virtual Private Networks (VPN)

VPN technology is the means of using a public infrastructure—the Internet—for the exchange of corporate information. VPNs use the Internet as a connectivity vehicle between telecommuters and the corporate environment. The primary difference between a simple Internet connection and a VPN is in the use of the public key infrastructure (PKI). The VPN makes use of the PKI to ensure a high degree of security between the users of the network. PKI also insures a high degree of security and privacy from any unwanted communications with a user outside the bounds of the established VPN.

Several security methodologies are currently being evaluated for VPNs. The most common is the secure internet protocol, often referred as *IPsec*. Ipsec is a standard internet protocol developed and standardized by the Internet Engineering Task Force (IETF). It is documented as a public request for comment (RFC) in anticipation of vendor support and implementation.

SOLUTIONS AND RECOMMENDATIONS

Analysis

As decision makers compile the enormous amount of information in this text, it becomes paramount to include the telecommunications infrastructure as a primary decision point. The following inclusive table (Table 1) condenses key information about the technologies discussed in this chapter, which are available from LECs throughout the United States and Canada. The information about the technologies is condensed into a comprehensive format to aid decision makers in selecting options for a telecommuting workforce. Table 1 also can offer the decision-makers a quick reference when discussing connectivity options with LECs, IXCs, and ISPs.

Future Trends

Even as this chapter is being written, new technologies are being developed that may have a far-reaching impact on future telecommuters. Since it is beyond the scope of this chapter to expound on future technologies, the following list of opportunities is provided for readers to explore.

Active Networks

Also called Programmable Networks, Active Networks came into probable reality through research funded by DARPA. In an article regarding Active Networks (Wetherall, 1998), the authors explore how the deployment of Active Nodes could use the current bandwidth of the Internet to increase overall network efficiency. The authors focus on a project constructed at MIT titled ANTS (Active Network Transport System). Several examples are used to illustrate the Active Network concept including stock quotes and on-line auctions. This may profoundly impact telecommuter productivity. The current telecommunications infrastructure can be maintained and still support LAN-like performance by incorporation Active Network technology.

Effortless Networking

In an article regarding Effortless Networking (Callahan, 1998), the author outlines the notion that "...device-independent, location-independent networking; enable peer-to-peer communications between devices; and let users instantly and automatically link to the network." Although the author spends significant time on *Jini* (Sun Microsystems, 1998), he does outline other technologies like *McoM* (Mobile Communicator/Communication), which is a wireless network being developed at Microsoft research, and *Salutation* (http://www.salutation.org), being developed by such corporations as Cisco and IBM. Although still in development, these technologies will impact the telecommuter from a Help Desk perspective. *IF* connectivity and configuration are now part of the network, the corporate Help Desk function may not have to expand to support the telecommuter work force.

Home Networks

An excellent article regarding home networks can be found in Lawton (1998). In this article, the author discusses the power of home computers and how they could easily become servers. The author also briefly discusses some current "standards" activities such as *HomeAPI*, *HAVi*, Microsoft's *Universal Plug-and-Play* and Sun Microsystem's *Jini,* all of which make home networking more of a reality than a topic of future research. These technologies further enhance the performance and productivity of the telecommuting workforce by allowing new workstations, network protocols and applications to integrate "effortlessly" into the home network environment.

CONCLUSIONS

When considering a telecommuting workforce, two possible situations are prevalent:
1. The telecommunications infrastructure is assumed to be a non-factor, thus the primary concern centers on the management and productivity of the telecommuting work force; or
2. The corporation considers the telecommunications infrastructure key to the success of the telecommuter workforce, and will invest in a technology search of possible alternatives as a way to maximize telecommuter productivity.

Table 1: A comprehensive overview of telecommunications technologies

Technology	Available to the Home?	Approximate Cost	Speed/Bandwidth	Network Management Available for Remote Support?	Requires Changes to the Corporate Infrastructure and/or Mission	Supported Protocols	QoS Attributes	Comments
PSTN	Yes, in most locations.	$15/month	Up to 52 Kb/s with special modems.	SNMP management only if SLIP or PPP used.	No	SLIP, PPP, TCP/IP	No guarantees.	POTS. Already available.
T-1/T-3	Yes, in most locations.	$100 to $500 per month based on service & distance.	1.544 Mb/s and ~45 Mb/s, with variations.	SNMP management only if SLIP or PPP used.	No	SLIP, PPP, TCP/IP	No guarantees	Dedicated line (trunk).
ISDN	Yes, in most locations.	$30 to $60 per month based on LEC/ISP + TA or Router.	From 64 Kb/s to 144 Kb/s	Supplied by LEC	No, as long as ISDN is at the corporate site.	SLIP, PPP, TCP/IP	No guarantees	Switched and dedicated.
Frame Relay	Yes, in most locations.	$20 to $50 per month + setup + report services (Interprise, 1999).	56 Kb/s or 1.544 Mb/s	Supplied by LEC	No, as long as Frame Relay is at the corporate site.	SLIP, PPP, TCP/IP	Crude. Frame Relay is a version of X.25, packet delivery can be specified.	A good starting point if LEC provides service to a private home.
ADSL	In some locations. Service is a function of copper cable service to the home.	$20 to $60/month + equipment + installation (i.e., $200 + $99) (OpNet, 1999) (Slip.Net, 1999).	Hundreds of Kb/s to 6 Mb/s depending on provider and cable distance.	Modem device checking functions. SNMP is optional.	Yes—Corporate IT will need an ADSL Gateway to the internal infrastructure.	Function of provider, but TCP/IP has been advertised.	No guarantees	Captive user of LEC. High cost to switch to a different Internet provider.
Cable Modem	In some locations. Service is a function of the cable TV ability to provide service.	$30 to $50 per month based on cable co. + modem (~$200).	Hundreds of Kb/s to 10 Mb/s depending on provider and number of users.	Cable TV Device checking functions. SNMP is optional.	Yes—Corporate IT will need a Cable TV Gateway to the internal infrastructure.	Function of provider, but TCP/IP has been advertised.	No guarantees	Captive user of cable company. High cost to switch to a different Internet provider.
FTTC	Usually in new home developments.	"Dark Fiber" pricing not available.	Could approach 10s of Gb/s rates.	A function of devices connected to the fiber.	Yes—a Gateway to the fiber plant.	N/A	N/A	This is a fiber plant. Dark Fiber is possible.
HFC	Possibly	F(speed)	Mb/s	CMISE	Gateway	TCP/IP	None	Cable TV
Wireless	In some locations	Unknown	Mb/s expected	No, or very crude	LMDS Gateway	N/A	No guarantees	New technology

In either case, corporations may have to invest and partner with service providers to build or rebuild the public infrastructure to meet the needs of the telecommuting work force. If corporations find themselves in this situation, a more in-depth review of the technologies previously presented may be needed to best use corporate investments.

ENDNOTES

1   This assumes an NIC card at $50 and a multiport Ethernet HUB costing about $20 per Ethernet port.

2   An Inter-LATA system is the equipment and plant owned and operated by the long-distance service provider. Examples include AT&T, Sprint, MCI, and GTE.

3   This assumes no compression and two cycles to represent one digital bit. Early modem technology conforming to these assumptions included rates of 300, 600, and 1,200 bits per second. They typically generated two tones to represent either a digital "1" or a digital "0."

4   Defense Advanced Research Projects Agency (DARPA) funded the design and deployment of he Internet. This started in the late 1970s and continues today.

5   Backbone is a phrase used to describe the main connecting technology to more subordinate network connections. The "trunk" term used by the telephone industry describes the 1.544 Mb/s backbones used to connect Toll Offices. Therefore, backbone and trunk are synonymous in many network topologies.

6   An acronym for its French name: Comite Consultatif Internationale Telegraphique et Telephonique. The CCITT is an international standards committee whose charter is to create standards for the international telecommunications industry. This group recently changed its name to International Telecommunications Union (ITU).

7   These rates assume that the B-Channel is available for data transfer. If not, then 2B+D = 128Kb/s and 23B+D = 1.372Mb/s.

REFERENCES

Bumblis, J. R. (1997). *The Enterprise Backbone: Making Sense of Current and Emerging Technologies*; ICC'97 Montreal, 1997 First IEEE Enterprise Networking Mini-Conference, Conference Record, 27-36.

Callahan, J. (1998). *Moving Toward Effortless Networking*, in the IEEE Computer Magazine, 12-14, November, 1998.

Doll, D. R. (1978). *Data Communications: Facilities, Networks, and Systems Design*; John Wiley & Sons, 1978, ISBN 0-471-21768-9.

Goldman, J. E. (1998a). *Applied Data Communications: A Business-Oriented Approach. (Second Ed).* New York: John Wiley & Sons, Inc.

Goldman, J. E. (1998b). *Applied Data Communications: A Business-Oriented Approach, Second Edition.* New York: John Wiley & Sons, Inc.

Jini (1998). Sun Microsystems Web page, *http://java.sun.com/products/jini/index.html.*

Lawton, G. (1998). New Technologies Take the Network Home, *IEEE Computer Magazine*, March 1999, 11-14.

Mahal, J. S. & Moore, D. (1999). Kwantlen University College, WWW Document; *http://www.geocities.com/CapeCanaveral/Lab/6800/P97_JDSR.HTM.*

McConnell, B. (1997a). Hello Direct, Inc., WWW Documents; http://*www.phonezone.com/tutorial.*

McConnell, B. (1997b). Hello Direct, Inc., December 1997, WWW Document; *http://www.phonezone.com/tutorial/iservice.htm.*

McConnell, B. (1997c). Hello Direct, Inc., December 1997, WWW Document; *http://www.phonezone.com/tutorial/analog-digital.htm.*

Noll, M. A. (1999). Does Data Traffic Exceed Voice Traffic?, *Communications of the ACM,* 42(6), 121-124.

pNet Home Page (1999). WWW Document; h*ttp://www.novusdesign.com/adsl/pricing.html.*

Pan, Y., Qiao, C, & Yang, Y. (1999). Optical Multistage Interconnection Networks: New Challenges and Approaches, *IEEE Communications*, February, 37(2), 50-56.

Ryan, J. P. (1998). WDM: North American Deployment Trends, *IEEE Communications*, February, 36(2), 40-44.

Slip.Net Information Page (1999). WWW Document; h*ttp://www.rodegard.com/dedicatedsales/adsl_pricing.html.*

Tanenbaum, A. S. (1988). *Computer Networks* (Second Ed), Englewood Cliffs, NJ: Prentice-Hall, Inc.

Wetherall, D. (1988*).* Introducing New Internet Services: Why and How, *IEEE Network Magazine*, May/June, 12-19.

U S WEST !NTERPRISE. (1999). Web Page, WWW Document; *http://www.inficad.com/~rfollow/frr.html.*

EMPLOYER
ISSUES

Chapter V

Success in The International Virtual Office

Kirk St. Amant
University of Minnesota, USA

ABSTRACT

The advent of technologies such as company e-mail systems and corporate intranets has expanded the office to include co-workers from various national and cultural backgrounds. This new development means that certain deep-seated cultural factors can affect interoffice communications in the new workplace. This essay presents some of the more prevalent cultural "problem" areas that can affect international and intercultural communication in the new virtual workplace. This essay also provides a series of tips or strategies that can facilitate effective intercultural communication in the new atmosphere of the virtual office area as well as overview certain resources that can facilitate communication in this international and multicultural environment.

INTRODUCTION

Revolutions in electronic communication are constantly changing how individuals think about "the office." The office was once a physical structure, and communicating with colleagues often meant wandering over to their cubicles to ask a question or to discuss an issue. Now, with technologies such as company e-mail systems and corporate intranets, the office has expanded to include co-workers from various national and cultural backgrounds and who are stationed in various locations around the globe. Together these individuals create a new kind of "virtual office" (a network of co-workers connected by nothing more than various on-line communication technologies) that is becoming increasingly common in the business world of the 21st century.

While colleagues in the virtual office might use similar technologies to communicate and often use a common language (primarily English), there are still deep-seated cultural factors that can affect the quality of interoffice communications in this new workplace. These differences often occur independent from technology or language, yet they can be among the most subtle and problematic factors contributing to confusion and miscommunication. For this reason, the following essay will not examine the technological differences that can exist among various regions or various nations. Rather, this essay will examine some of the more prevalent cultural "problem" areas that can affect communication in the new virtual office. Also, as many individuals new to intercultural communication might not be aware that certain cultural differences exist, the following essay focuses on heightening reader awareness of these issues and on providing strategies for overcoming these problems rather than engaging in a deep, academic examination of these problem areas.

LITERATURE REVIEW/
SITUATIONAL OVERVIEW

Different cultures have unique expectations concerning information presentation, and these can vary depending on the specific cultures represented and on the topic of the interaction. For example, individuals from different cultures often structure messages differently despite similar linguistic backgrounds (e.g., British and New Zealand English vs. US English) (Driskill, 1996). Similarly, individuals from different cultures often use different strategies for proving a point or for persuading an audience. In fact, manuscripts written by German or French authors often present broad conclusions unsupported by data while British and American authors usually use extensive data analysis but present few conclusions (Hofstede, 1997).

Cultural expectations can also occur at a more micro level, for different cultures can have different expectations of sentence length, and these expectations can affect how individuals from a certain culture perceive the importance or the credibility of a given message. As Ulijn and Strother (1995) have noted, certain southern-European clients "might believe that the product is not a serious one because the sentences in the instructions telling how to use the product are too short," and Li and Koole (1998) have shown that these cultural differences can even cause intercultural communication problems at the word level. Moreover, one study indicates that an individual judges a message according to the communication expectations of his or her native culture, even when that message is written in another language (Ulijn, 1996). However, there is some evidence that the more one is exposed to another culture, the better one gets at understanding how to design messages for members of that culture (Campbell, 1998). And while little has been written on the effects on-line media are having on intercultural exchanges, there is some evidence that cultural differences in message design persist even with younger individuals who are more "cyberspace savvy" (Ulijn & Campbell, 1999), possibly because individuals in different cultures tend to access and use the on-line environment differently (Gillette, 1999). Thus certain cultural communication preferences can lead to misunderstanding in the new area of on-line communication, and some of the more problematic areas will be examined in this essay.

LANGUAGE OF COMMUNICATION

The fact that an individual from another culture "speaks" English does not necessarily mean that individual speaks English well or that the speaker understands all of the subtle nuances and intricate uses of the language (Varner & Beamer, 1995). Certain geopolitical and economic factors have greatly affected the nature of English language education in other countries. Factors such as limited access to native English speakers, limited access to good teaching materials and effective and competent instructors, or the nature of a given educational system can all impact a non-native speaker's ability to comprehend and create English-language communiqués (Rodman, 1996; Katzenstein, 1989). Even for those individuals who have studied English for a number of years under an effective educational system, certain linguistic aspects could contribute to cross-cultural confusion when English is the language of the exchange (Murdick, 1999). Moreover, there are different dialects of English used around the world, and while often quite similar, they are still different enough to result in communication problems (Crystal, 1995).

By using certain strategies, individuals in the virtual office can create "user-friendly" English-language communiqués for an international audience. First, *avoid idiomatic expressions* (word combinations that have a specific cultural meaning different from their literal meaning—e.g., "It is raining cats and dogs" does not literally mean that cats and dogs are falling from the sky like raindrops). Second, *avoid abbreviations* (like idioms, they rely on a particular cultural background to understand what they mean). If abbreviating is essential to maintaining specific page-length requirements, spell out the actual term the first time the abbreviation is used and then use some special indicator to demonstrate how this abbreviation relates to the original term or expression (e.g., "This passage examines the role of the *Internal Revenue Service* (IRS).") Individuals also need to *identify and avoid false cognates* (terms in one language that sound like and look like terms in another but have different meanings—e.g., the word "map" exists in both Dutch and English, but in Dutch "map" means "folder" (Ulijn & Strother, 1995). Perhaps the best way to avoid using false cognates is to have a native from the target audience review the final English-language document, identify the false cognates and remove or replace them. Workers in the virtual office should also a*void using complex verbs or long noun strings.* (If more than one verb is used in a sentence, speakers with a limited knowledge of English might have problems discerning what the actual verb of the sentence is; similarly, long noun strings could make it difficult for these individuals to identify the actual subjects of sentences—e.g., "A light particle flux time series is the subject of this paper" [Jones, 1996]). Finally, when using English in the virtual office, workers also need to *know the dialect of English spoken by the target culture.* Speakers of the various dialects of English could have different terms for the same object or concept (e.g., American mechanics use "wrenches" to change a car's "battery," but their British counterparts use "spanners" to change "accumulators") or could have different meanings for the same term (e.g., in British English, a "subway" is a highway underpass, while in American English, a "subway" is an underground train system). Various dialects of English can also have different idiomatic expressions or different spellings for the same word (e.g., "color" vs. "colour"). Thus, employees in the virtual office need to identify the dialect of English spoken by their international target audience and either familiarize themselves with potentially troublesome expressions or have a native speaker of that dialect of English review the document to help ensure that the correct message is conveyed.

NUMERIC REPRESENTATION

Numbers are one of the key components of most business communications; however, as with the English language, different cultures have different ways of representing the same idea, such as dates, times, and magnitudes. Failure to understand and recognize these differences can lead to embarrassment, confusion, or costly mistakes. Employees in the international virtual office, therefore, need to understand how culture can affect numeric representation if they wish to present or understand important numeric information.

In the United States, dates tend to be written in the order of the month, the day, and then the year (e.g., April 1, 1998 becomes 4-1-98). In many European countries, however, the order of the month and the day are reversed (1 April 1998 becomes 1-4-98), and these differences can result in problems if members of these two cultures use numbers to signify a meeting date, a key production deadline, or a shipping date. To avoid this problem, spell out the name of the month when presenting dates in a document. For example, dates written as January 1, 1999 and 1 janvier 1999 avoid month/day confusion.

Written representations of time can also vary from culture to culture. In the United States, for example, most individuals tell time according to a twelve-hour clock. Many other cultures, however, tell time based on a twenty-four-hour clock. In these cultures, p.m. times are related by adding twelve to the given hour. For example, 1:00 p.m. becomes 13:00. Other cultures, such as the Quebecois, use numeric representations that outsiders might not recognize as representing time (e.g., 9:40 p.m. US time becomes 21h40 Quebecois time). As with dates, the various cultural numeric representations of time can cause confusion or result in readers misinterpreting the significance of a particular piece of numeric data. To avoid this problem, let the reader know what system of time representation is used in the document (e.g., "This document relates time based on a 24-hour clock, with which 0:00 represents midnight."). This statement should appear at the beginning of a document and at the beginning of every pertinent section. Also, employees should try to use a 24-hour clock to avoid confusion concerning a.m. and p.m.

Many cultures have also developed a series of punctuation-based markers that help readers recognize magnitude. In the United States for example, commas are routinely used to denote numeric magnitude for thousands (1,000), millions (1,000,000), billions (1,000,000,000), and so forth. A German co-worker, however, would most likely use a period to represent magnitudes such as one thousand (1.000) or one million (1.000.000). Moreover, the Swedes often use a space rather than a punctuation mark to represent magnitudes (one thousand is 1 000 and one million is 1 000 000).

Another magnitude problem is that different cultures might use the same punctuation marks to indicate different scales of magnitude. Americans use commas to indicate one thousand or greater and periods to indicate values smaller than one. In this system, one thousand five hundred is written as 1,500, while one and one half is written as 1.500. The French, however, use the reverse of this system of punctuation. For example, the French represent one thousand five hundred as 1.500 and write one and one half as 1,500. Such differences could easily cause confusion or misinterpretation, especially if the reader is from another culture and is also in a hurry.

To avoid these problems, first know the magnitude format used by that audience and present all magnitude data in that fashion. Next, begin a communiqué by clearly and explicitly stating what system of magnitude representation the communiqué will use, and be sure to discuss orders of relative magnitude (e.g., 1,500 = one thousand five hundred, and 1.500 = one and one half). Finally, restate the system of magnitude representation at the beginning of each section containing magnitudes to ensure that the reader will correctly interpret the information.

International communication can also often involve two different systems of measurement: the English system (inches, feet, yards, and miles) used mainly by the United States, and the metric system (centimeters, meters, and kilometers) used by the rest of the world. Failure to understand this difference could lead to confusion if individuals from one culture receive data in a measurement system they are unfamiliar with and do not know how to convert these items to their own measurement system. To avoid these problems, mention the system of measurement being used at the beginning of each message. If possible, include exact measurements in both English and metric systems (e.g., move the item 1 inch/2.54 centimeters). Also, if the document uses only one system of measurement, be sure to include a measurement conversion table to help readers correctly understand measurements and values.

MAKING CONTACT

Operating effectively in the virtual office requires one key factor: contact. With ubiquitous e-mail and cellular telephones, individuals from certain cultures expect e-mails and voicemails to be answered quickly and for someone to always be "on call" at a particular office. However, not all cultures share this same perception of time, and this difference in perception can affect how "accessible" individuals from other cultures are.

While many Americans feel the compulsion to return e-mail quickly, not all individuals share this practice. In Ukrainian culture, for example, face-to-face communication tends to be valued over other forms of interaction, especially in a business setting which could explain why days and even weeks can go by before e-mails to the Ukraine receive a response (Richmond, 1995). This long "turnaround time" appears to be, in part, due to cultural factors, for many individuals from other cultures do not see the need to respond to e-mail with the same degree of urgency that most Americans do (V. Mikelonis, personal communication, Aug. 20, 1999). Similarly, in certain cultures, the importance of a message is often linked to the amount of effort that went in to delivering the message. In parts of Kenya, for example, messages delivered by a person who walked across town and up three flights of stairs to deliver the message in person (great expenditure of personal effort) are often seen as more important and got a quicker response than e-mail messages that only required the press of a button to send (low expenditure of personal effort) (Van Ryckeghem, 1995; N. Johnson, personal communication, Feb. 20, 2000).

The time at which one attempts to contact other locations could also account for some of these problems. Americans, for example, are accustomed to contacting places of business at essentially any time during the day or during the year (with the exception of certain national holidays such as Thanksgiving or Christmas). Moreover, even in the thick of vacation season, the average American expects to find some staff working at a place of business, especially if that business is a large, multinational company. However, in Spain or Italy (especially in the regions outside of major urban areas), it is not uncommon for an office to shut down for two or more hours in the middle of the day to observe the traditional two-hour lunch period (generally from noon to 2:00 p.m. or from 1:00 - 3:00 p.m.) (D. St. Amant, personal communication, Aug. 30, 1999). Similarly, many Americans think of the workweek as beginning on Monday and ending on Friday. In many Islamic countries, however, it is not uncommon for business to be closed for part of or for all of the day on Friday (the Islamic holy day) but to be open on Saturday or Sunday (N. Johnson, personal communication, Feb. 20, 2000). Also, most Americans think of vacations in term of 2 or 3 weeks, and even during the height of the summer vacation season, there is often someone in the office to answer the phones. The French, however, take their summer vacations more seriously, and it is not uncommon for businesses, if not entire towns, to "shut down for the summer" while all of the employees (or residents) go away on vacation. (As Weiss [1998] puts it, "[O]ne is hard pressed to conduct business [in France] between July 14 and September 1"). And different cultures could also observe different holiday vacation periods (e.g., the Japanese "Golden Week" from April 29 to May 5) (Holroyd & Coates, 1999).

International time differences can also make quick responses to unscheduled message difficult, as one group of workers could be leaving work at the same time virtual co-workers in other parts of the world are arriving at work. Furthermore, the notions of "today," "yesterday," and "tomorrow" can become very confusing depending on the time difference between the sender and the receiver of a message. If, for example, an individual in the US wishes to tell a colleague in the Netherlands that a report is needed by "tomorrow," does the person in the US mean tomorrow according to US time (which could easily be today Netherlands time, depending on when the message was sent) or does that colleague mean tomorrow according to Netherlands time (which could easily be two days from the time at which the person in the US sent the message)?

By learning about other cultures, one can increase the chances of "making contact" by following a few simple steps. First, schedule times of important information transfer well in advance to ensure that the needed parties will be present at the needed time. Second, prearrange the best means of contacting others when a quick response is essential, and then determine a necessary backup plan if this option fails. Next, alter schedules to accommodate holidays and vacation periods for all parties. Also, know the time differences for overseas offices that are regularly contacted to better understand when the earliest response to a given message could occur. (Certain scheduling software, such as Microsoft Outlook, allow for scheduling in more than one time zone while other kinds of scheduling software let users insert international holidays into their calendars). Finally, avoid relative date references (e.g., "tomorrow" or "yesterday"). Instead, give the day and the date ("Monday, October 4") as well as some chronological context according to the recipient's time frame (e.g., "Netherlands time"). By following these steps, the employees in the virtual office should be able to increase the chances of making contact with co-workers in various parts of the world.

CULTURAL APPRECIATION

In many cases of intercultural communication, there is often a sense of suspicion associated with "outsiders." Therefore, it is wise to establish good working relations with individuals from other cultures early on in a business relationship. One of the simplest and easiest ways to achieve such relations is to display an understanding of and an appreciation for the cultures with which one is dealing. For example, during a 1980 exchange of American writers to Poland, Susan Sontag quickly developed a good rapport with her Polish hosts and became "the darling" of the exchange because she had familiarized herself with Polish history and culture in the weeks prior to her visit. As a result, Sontag could show her hosts that she recognized and appreciated their culture, and this cultural knowledge expanded the range of topics she could discuss with the Poles (Richmond, 1995).

Displaying such an appreciation for the "other" culture is important on two key levels. First, it creates a positive first impression, for it shows that the individual has taken the time to do some research on the culture or the company (responsibility and initiative) and implies that the person is open to working with individuals from other cultures (objectivity). Second, displaying knowledge of the other culture can indicate a degree of commitment to the overall relationship. Thus, employees in the virtual office should demonstrate knowledge of the other culture at the time of initial contact.

There are two key ways in which the individual can display an appreciation for the cultural background of other persons. First, capitalize on knowledge of geographic information. For example, if a Dutch counterpart says he or she is from Eindhoven, the virtual office mate should learn where the town is, and then follow up with a question such as, "Eindhoven is in the southwestern part of the Netherlands, is it not?" or "Eindhoven is next to Maastricht, isn't it?" Second, try to use key phrases (e.g., opening and closing statements) in the native language of the overseas counterpart, especially during early interactions. These simple opening and closing phrases (e.g., "hello," "how are you" or "thank you" or "have a good day") show an appreciation for another culture and display an openness to working with other linguistic groups. Using phrases in another language has the benefit of being an "offensive" strategy—one doesn't have to wait for members of the other culture to provide the geographic or other cultural information to present a culturally savvy response. Also, these phrases do not necessarily need to be memorized. Rather, one could purchase a basic foreign language phrase book and use it to voice different expressions in other languages. Individuals should, however, stick to the basic phrases, for attempting to devise more complex communications in another language could result in awkward and embarrassing messages. These two strategies have the added benefit of being rather unobtrusive and can be used regularly and over long periods of time without becoming old.

FORMALITY

Different cultures expect a particular level of formality to be maintained during the course of an interaction, especially in a business context. And this notion of formality carries over into the way in which members of different cultures draft business communications. French business communication, for example, tends to be very formal. One French composition text encourage readers to begin business letters with phrases such as, "J'ai le plaisir de poser ma candiature au poste de . . ." (I have the pleasure of posting my candidature for the position of . . .). The same text encourages readers to end business letters with equally formal phrases such as, " . . . l'assurance de mes sentiments les plus distingués" (. . . the assurance of my most distinguished sentiments) (Simard, 1998).

Other cultures, however, tend to have less rigid expectations of formality, especially in relation to on-line communication. Americans, for example, might find the writing style of a French letter overly formal, wordy, or even stilted. However, French readers who have been instructed to write in this format might consider letters written in a less formal style not credible or even rude or insulting. Moreover, these expectations of formality seem to carry over into the on-line environment (E. Watts, personal communication, Sept. 20, 1999).

Cultural expectations of formality also have important implications for name usage: specifically, when can an individual address someone from another culture by first name. In US business correspondence, for example, it is not uncommon to address someone by his or her first (given) name after three or four correspondences. Germans, however, often consider the use of first names in business correspondences "inappropriate chumminess" or even "rudeness." In fact, for many Germans, friendliness, trust and respect are often displayed through the use of both titles and family names/surnames, e.g., Dear Director Schultz (Bell & Williams, 1999).

Furthermore, different cultures have different ways for organizing an individual's first/given name and family name/surname. In the US, for example, the given name comes first and the family name comes second (e.g., John Smith). In many Asian cultures, however, the family name comes first and the given name comes second (e.g., Smith John). Alternatively, in many Latin American cultures, names include both the father and the mother's family names as well as the given name, so a person's full name is presented as first name, father's family name, mother's family name (e.g., John Smith Jones), and it is the father's family name that is used in formal addresses with titles (e.g., "John Smith Jones" becomes "Señor Smith"). Thus, individuals wishing to maintain the correct level of formality must make sure they are aware of a given culture's naming system to give the correct, formal address (e.g., "Mr. Smith" vs. "Mr. John").

To avoid potential insult related to expectations of formality, one can adopt three key strategies. First, use a formal tone when communicating with individuals from other cultures, especially when communicating with new acquaintances or superiors. Second, let the person from the other culture determine when the tone of the business relationship can become more informal. Finally, avoid given names when addressing someone from another culture. Rather, use titles ("Mr." or "Ms." or "Dr.") and family names when addressing someone for the first time and do not use an individual's first name until that individual gives permission to do so (in some cultures, it is permissible for a superior to refer to a subordinate by his or her first name while that subordinate should not refer to the superior by first name). Ideally, employees in the virtual office would learn the titles of respect used in a given culture and then use those titles when communicating with persons from that culture (e.g., Herr Schmit, Tokida-san, Madame St. Onge).

STATUS AND DISTANCE

In some cultures, individuals can sometimes circumvent official channels to achieve a particular goal. In other cultures, however, the corporate hierarchy is rigid, and employees must go through the expected formal channels to get results. In such systems, attempts to circumvent the standard process are looked upon unfavorably and could damage the reputation of individuals who use such methods. Also, such a hierarchy breach could be something as simple as sending an e-mail to someone at a higher corporate level (a practice that is not necessarily uncommon in the United States), and such attempts could either be futile (the individual simply ignores the message) or have detrimental effects (the individual becomes angered by such brashness and decides to punish the sender for insubordination).

The sociologist Hofstede (1997) dubbed this notion of how adamantly different cultures adhered to a hierarchical system as "power distance," and he created a chart to compare and contrast cultural power distance relations.

One should keep in mind that different kinds of power distance relations can exist within a given culture and can be based on social class, ethnicity, religion, gender, or age.

Table 1: An overview of cultural perspectives of power distance

High Degree of	Panama
Power Distance	Mexico
	Arab Countries
	France
	Hong Kong
	Taiwan
	Japan
	USA
	Germany
Low Degree of	Great Britain
Power Distance	Sweden

To avoid turmoil within a given cultural system, employees in the virtual office should proceed with extreme care when dealing with corporate hierarchies in other cultures. These individuals should first learn the hierarchy structure present in a certain culture and learn how closely individuals must follow that system as well as the repercussions that could result from circumventing the system. Next, individuals should determine who their equivalents are in the context of another corporate culture (to ensure that messages get sent to the correct individual and not to someone at a higher point in the power structure) and attempt to work within this power structure when seeking or sending information. Finally, workers in the virtual office should check with counterparts in a particular culture to make sure it is acceptable to contact individuals in higher positions within that power structure.

DIRECTNESS

The purpose of a business communication is to exchange information. However, the way in which that information is presented can vary greatly between cultures, and such variations can cause confusion and frustration. One of the most prevalent problems related to intercultural information exchange involves directness or explicitness. For example, most Americans get to the point quickly and directly state the purpose of a message. Moreover, Americans often consider indirectness as suspicious behavior, for an "honest" or "competent" person would get to the point right away. For many

Japanese communicators, however, indirectness is key to an effective business exchange. As Murdick (1999) puts it, "The Japanese prefer an indirect rhetorical approach in which the writer approaches the main point slowly and perhaps never directly states it but merely suggests or hints at the intended message," for in Japan, "direct questioning about the obvious is considered childish."

These different attitudes concerning directness can cause problems. For example, direct American presentations can offend a Japanese individual who might think that he or she is being "talked down to" (Murdick, 1999). Conversely, an American reader could consider an indirect Japanese business memo dishonest (the writer is "beating around the bush"). Thus, an understanding of how different cultures perceive and use directness and indirectness is essential to effective communication in the virtual office, and the areas in which such differences can cause the greatest problems (saying "no" and publicly expressing displeasure) both relate to the preservation of "face" or external public image.

In some cultures, directly saying "no" can cause the speaker or the hearer to lose face (create a poor public image). In these cultures, "no" is often expressed indirectly with phrases such as "I will have to get back to you on that" or "I think so, but I am not sure for certain" (M. Mendez, personal communication, Jan. 14, 2000). Such indirect expressions can cause two kinds of intercultural communication problems. First, individuals from more direct cultures might not realize that such answers mean "no" and thus, these more direct individuals could either wait for an answer that never comes (" I will have to get back to you on that") or mistakenly assume that a "yes" answer was given ("I think so, but I am not sure for certain"). Second, individuals from more direct cultures could accidentally give an answer ("I will have to get back to you on that") that someone from a less direct culture could misinterpret as meaning "no," resulting in the opposite of what was intended.

To avoid such problems, individuals from more direct cultures should use questions that let the other party provide an answer other than "no." Thus, instead of asking, "Will the shipment arrive by May 30?" the question should be, "When do you think the shipment will arrive?" (The second question uses a structure that lets the asker know if the May 30 deadline will be met while allowing the other party to avoid giving a direct "no" response.)

The direct expression of displeasure or dissatisfaction (e.g., "I dislike this product or service") can also cause a loss of face in certain indirect cultures. In such cultures, displeasure is often expressed through omission. For example, if an individual receives two items and likes one but not the other, that individual will praise the quality of the item he or she liked, but say nothing at all about the item that he or she disliked, and this omission (saying nothing) would indicate displeasure or disapproval. Unfortunately, individuals from more direct cultures might not realize that, in some cases, failure to discuss an item indicates displeasure concerning that item, which could result in individuals from the more direct culture repeating the same offense. Thus, workers in the virtual office need to know when to be direct or indirect in voicing displeasure and when to interpret an omission as an indirect expression of displeasure.

While there is no simple way to identify or to understand all of the intricate communicative styles of implicit cultures, Hall has devised a *context* model that can help workers in the virtual office gain a basic understanding of how different cultures might perceive implicit or explicit messages. In this model, individuals from *high-context* cultures tend to be more indirect when presenting information and often use silence as a way to convey meaning. Conversely, individuals from *low-context* cultures expect information to be presented directly and explicitly.

Understanding these context-based differences can help individuals avoid presenting too little or too much information, depending on the cultural audience, and can prevent incorrect assumptions about co-workers from other cultures (e.g., this person must be shifty, or he/she would get to the point).

While dealing with the directness/indirectness differences is not simple, the following steps can help employees in the virtual office create more

Table 2: An overview of Hall's ideas of culture and context (Hoft, 1995)

High-context Culture (Prefers information be stated *implicitly*)	Japanese Arabic nations Latin American cultures Italians English French North Americans (USA and Canada) Scandinavian cultures (Norway,
Low-context Culture (Prefers information be stated *explicitly*)	Denmark, Sweden) German Swiss-German

effective communiqués based on cultural expectations. First, individuals can use Hall's context model to get a general overview of a given culture's expectations related to explicit vs. implicit information presentation. Workers in the virtual office can then use this understanding of context for both drafting messages for and reading messages from individuals from other cultures. However, the best solution is to have a member of a particular culture read key memos or letters written by individuals from that culture. Such "native interpretation" can help ensure that the non-natives get the correct interpretation of a message, especially if the message relates to a key or important subject. (Similarly, these natives can review drafts of communiqués to determine if they are too explicit or too implicit for a given cultural audience.)

RESOURCES FOR EMPLOYEES

The best way for an individual employee to work effectively with individuals from another culture is to know something about that culture. The following cultural references are all inexpensive and easy to obtain or to access, and all of these resources allow individual employees to locate and access information quickly.

Culturgrams, published yearly by Brigham Young University, are four-page pamphlets/brochures that provide a quick overview of key aspects of a particular country (e.g., History, Language, General Attitudes, Government, Education), and nominal fee. They can be ordered toll free by calling 1-800-528-6279. (Note that not all nations will have a corresponding *Culturgram;* in these cases, individuals will need to seek alternative sources.)

The CIA World Factbook Online is a free-access, on-line governmental site containing easy-to-access current and specific information on individual countries. Located at http://www.odci.gov/cia/publications/factbook, this site contains more in-depth information than that found in the Culturgrams. It also contains a search option that helps users locate specific information on a given country.

The *Online UN Embassy Site* is free to use and provides ready access to several of the web sites of member nation embassies. The overall embassy site is located at http://www.un.org/Overview/unmember.html and provides access to specific embassy web sites. The various embassy web sites can differ in the extent of and kind of information they have on their particular countries, but many of these sites have an e-mail address that allows the user to direct culture-specific questions to embassy staffers.

The *US Department of State Electronic Research Collection* is a free-access on-line site that allows users to search various State Department databases for particular information relating to certain countries. The database is located at http://dosfan.lib.uic.edu/ and contains much more in-depth, country-specific information than the *CIA Factbook* or the *Culturgrams*. Users should check the publication/release dates of documents to make sure they are relatively current and keep in mind that some of the documents in these databases are relatively lengthy.

Finally, *natives from a given target culture* can provide valuable information on the communication expectations of their fellow countrypersons as well as serve as a sample audience to determine if English-language documents are suited to the linguistic skills of a given cultural audience. However, one needs to determine if the natives with which he or she is working are actually accurate representatives of a larger cultural audience (i.e., is the subject's level of English typical of someone from that culture, or has that subject received any special English-language training or had more exposure to English or to English speakers than the average English-speaker in that country). Also important is the amount of time the native has been away from the target culture, for the more time spent away from that culture, the greater the chances are the culture has changed and that the information provided by this individual will be dated and possibly inaccurate.

For those individuals who want a more in-depth understanding of culture and communication, the following books provide a good, academic foundation of cultural expectations and how these expectations can affect cross-cultural communication:

The Cultural Context in Business Communication, Edited by Susanne Niemeier, Charles P. Campbell, and Rene Dirven, published by the John Benjamin Publishing Company (1998).

Culture and Organizations: Software of the Mind, by Geert Hofstede, published by McGraw-Hill (1997).

International Dimensions of Technical Communication, Edited by Deborah C. Andrews, published by the Society for Technical Communication (1996).

Managing Global Communication in Science and Technology, Edited by Peter J. Hager and H. J. Scheiber, published by John Wiley & Sons, Inc. (2000).

Intercultural Communication in the Global Workplace, by Iris Varner and Linda Beamer, published by Irwin/McGraw-Hill (1995).

These books provide a good deal of in-depth information concerning culture and communication as well as identify and explain various culture-based differences in communication styles and expectations. Also, many of the titles provide essays by different authors from different cultures, so the reader can get a more multicultural perspective on intercultural communication. Additionally, readers can use the "References" sections of these books to locate a wider variety of materials on intercultural communication issues.

SUGGESTIONS FOR MANAGERS

The key to a successful virtual office is knowledge, for the more employees know about the cultures of their international co-workers, the easier it is to avoid culture-based communication problems. Managers wishing to create and maintain an effective and efficient workplace, therefore should develop a strategy or a system for making their employees aware of these cultural differences and provide materials that will help employees communicate effectively in this environment. These seven steps provide various foundational strategies that managers can use to create a more effective intercultural communication environment.

1. *Identify* the cultures with which employees deal most often and design awareness training seminars focusing on that culture. Such seminars would overview geography, communication patterns (in terms of the problem areas mentioned earlier in this essay), history of the region/culture and an overview of the politics and social dynamic of that culture (factors that provide both a deeper understanding of the culture and a wider range of "discussion" topics to use in interactions).

2. *Provide* employees with contact information (e-mail address and phone or fax numbers) on their counterparts in other cultures and encourage them to interact with these counterparts to build relationships that the employee can then use as a foundation for asking culture-specific questions.

3. *Establish* a library of intercultural references with quick reference sources as well as more in-depth academic/scholarly articles and books on intercultural communication so employees have the opportunity to both quickly access the information they need and learn about the deeper factors involved in intercultural communication.

4. *Send* employees reminders about upcoming holidays or vacation periods related to the cultures with which they have contact; in some cases,

managers can use certain scheduling software (e.g., GroupWise) to send out mass notices that can be automatically added to employees' electronic calendars.

5. *Develop* a backup system for reaching international co-workers during vacation and holiday periods or in cases of technology failure.

6. *Remain* current by having employees review professional communication journals (e.g., *Technical Communication, Intercom, IEEE Transactions on Professional Communication*) and then write up and distribute summaries of articles on international or intercultural communication.

7. *Design* culture-specific, "quick reference" resources employees can use at their desks. Reference resources should include information such as major holidays/vacation periods, time differences, numeric information (how the culture represents times, dates, and magnitudes), maps of the given country, key basic phrases in the language of that culture, tips for writing for non-native English speakers (best designed with the help of someone who has training in ESL education and who has an understanding of the culture with which the employee is communicating), and suggestions for how formal/informal to be in communications (e.g., "Always begin by addressing individuals as . . .").

This list provides managers with a base on which they can build more in-depth intercultural communication training sessions and design more in-depth or culture-specific reference or training materials. To develop these training sessions and materials, managers might wish to contact other professionals who have successfully dealt with a given culture over time or work with individuals from a given culture to develop culture-specific materials. What is key is that the resulting materials are current and that they accurately reflect the cultural communication expectations of the individuals with which employees will be interacting. Individuals wanting more information on developing or designing materials, designing training programs, or persons who would like to have someone give a seminar for employees or managers can contact the author of this chapter via e-mail at stam0032@tc.umn.edu.

CONCLUSION

The rapid evolution of communication technology is constantly changing how people think about space. The relatively widespread use of e-mail and corporate intranets has begun to change the concept of "the office" from a physical location to a state of mind. As this virtual office continues to include increasing numbers of co-workers from different cultures and locations, the nature of future interoffice communication will almost certainly be shaped by intercultural communication issues. While this essay is by no means a comprehensive source on intercultural business communication, it does provide the reader with the basic understandings and strategies needed to function efficiently in the new international working environment. The reader must now take the next step and learn more about the cultures with which he or she will regularly interact, for only through such an understanding of other cultures can the international virtual office become a comfortable environment in which to work.

REFERENCES

Bell, A. H., & Williams, G. G. (1999). *Intercultural business*. Hauppauge, NY: Barons.

Campbell, C. P. (1998). Rhetorical ethos: A bridge between high-context and low-context cultures? In S. Niemeier, C. P. Campbell, & R. Dirven (Eds.), *The cultural context in business communication* (pp. 31-47). Philadelphia, PA: John Benjamin Publishing Company.

Crystal, D. (1995). *The Cambridge encyclopedia of the English language*. New York: Cambridge University Press.

Driskill, L. (1996). Collaborating across national and cultural borders. In D. C. Andrews (Ed.), *International dimensions of technical communication* (pp. 23-44). Arlington, VA: Society for Technical Communication.

Gillette, D. (1999). Web design for international audiences. *Intercom, December* 15-17.

Hofstede, G. (1997). *Cultures and organizations: Software of the mind*. New York: McGraw Hill.

Hoft, N. L (1995). *International technical communication: How to export information about high technology*. New York: Wiley.

Holroyd, C., & Coates, K. (1999). *Culture shock: Success secrets to maximize business in Japan*. Portland, OR: Graphic Arts Center.

Jones, A. R. (1996). Tips on preparing documents for translation. *Global Talk: Newsletter for the International Technical Communication SIG, 4*, 682, 693.

Katzenstein, G. J. *Funny business: An outsider's year in Japan.* New York: Soho.

Li, X., & Koole T. (1998). Cultural keywords in Chinese-Dutch business negotiations. In S. Niemeier, C. P. Campbell, & R. Dirven (Eds.), *The cultural context in business communication* (pp. 186-213). Philadelphia, PA: John Benjamin Publishing Company.

Murdick, W. (1999). *The portable business writer.* New York: Houghton Mifflin.

Richmond, Y. (1995). *From da to yes: Understanding the East Europeans.* Yarmouth, ME: Intercultural Press, Inc.

Rodman, L. (1996). Finding new communication paradigms for a new nation: Latvia. In D. C. Andrews (Ed.), *International dimensions of technical communication* (pp. 111-121). Arlington, VA: Society for Technical Communication.

Simard, J. P. (1998). *Guide du savoir-ecrire [Writing guide].* Montreal: Les Editions de l'Homme.

Ulijn, J. M. (1996). Translating the culture of technical documents: Some experimental evidence. In D. C. Andrews (Ed.), *International dimensions of technical communication* (pp. 69-86). Arlington, VA: Society for Technical Communication.

Ulijn, J. M., & Campbell, C. P. (1999). Technical innovations in communication: How to relate technology to business by a culturally reliable human interface. In *Proceedings of the 1999 IEEE international professional communication conference.* Piscataway, NJ: Institute of Electrical and Electronic Engineers, Inc.

Ulijn, J. M., & Strother J. B. (1995). *Communicating in business and technology: From psycholinguistic theory to international practice.* Frankfurt, Germany: Peter Lang.

Van Ryckeghem, D. (1995). Information technology in Kenya: A dynamic approach. *Telematics and Informatics, 12,* 57-65.

Varner, I., & Beamer, L. (1995). *Intercultural communication in the global workplace.* Boston: Irwin.

Weiss, S. E. (1998). Negotiating with foreign business persons: An introduction for Americans with propositions on six cultures. In S. Niemeier, C. P. Campbell, & R. Dirven (Eds.), *The cultural context in business communication* (pp. 51-118). Philadelphia, PA: John Benjamin Publishing Company.

Chapter VI

Organizational Compatibility as a Predictor of Telecommuting

Susan J. Harrington
Georgia College & State University, USA

Cynthia P. Ruppel
The University of Toledo, USA

ABSTRACT

Innovation literature long has advocated that an innovation may be compatible or incompatible with an organization's existing systems or resources. Compatibility of an innovation traditionally has meant that the innovation is compatible with the existing values, skills, and work practices of potential adopters. However, Tornatzky and Klein (1982) criticized this definition as too broad, noting that compatibility may refer to compatibility with the adopters' values (value compatibility) or that it may represent congruence with the adopters' existing practices (practical compatibility). Anecdotal evidence suggests both types influence telecommuting. Therefore, this study investigates compatibility and its relationship to IS personnel's telecommuting. The organization's ability to secure telecommuting (a dimension of practical compatibility) was found to be a major facilitator of the adoption and diffusion of telecommuting. Group values were found to be a major facilitator of diffusion and success, and practical compatibility was found to be a facilitator of success. Implications are discussed.

Predictions concerning the growth of telecommuting[2] have not material-ized, despite potential benefits to both employers and employees (Guthrie, 1997). Though reasons for this lack of growth are not clear, lack of control of telecommuting employees has been cited as one reason in that, while out of sight, employees will engage in opportunistic behavior (Bresnahan, 1998; Christensen, 1992; Handy, 1995). In a 1995 study, Hewitt Associates found that 63% of companies felt that a major drawback of telecommuting arrange-ments was reduced control and supervision by managers (Jones, 1996). Such arrangements characteristically lack the same type of controls present in traditional on-site work arrangements.

Success in telecommuting arrangements could be improved by using outcomes-based measures of employee performance, rather than manage-ment by observation or by whether or not the employee looks busy (DiMartino & Wirth, 1990). Thus, it appears that organizations that manage by setting objectives and by enhancing management and employee work skills would be those most compatible with telecommuting arrangements. Recently, some researchers (i.e., Belanger & Collins, 1998; Kavan & Saunders, 1998) have proposed that the compatibility of telecommuting, or other similar alternative work arrangements with the organization, individual, work, and technology, is important to the work arrangement's success.

The innovation literature long has advocated that an innovation may be compatible or incompatible with an organization's existing systems or resources (Rogers, 1983). Innovations compatible with existing resources imply that the risk of failure in implementing the innovation is reduced. The level of compatibility likely will be seen by the organization as a need for either a major reorientation or merely an adaptation (Downs & Mohr, 1976).

Compatibility of an innovation has traditionally meant that the innova-tion is compatible with the existing values, skills, and work practices of potential adopters. However, Tornatzky and Klein (1982), in a meta-analysis of innovation adoption and implementation, criticized the dual definition of compatibility as too broad, noting that compatibility may refer to compatibil-ity with the values or norms of potential adopters (value compatibility), or that it may represent congruence with existing practices of the adopters (practical compatibility). Similarly, Klein and Sorra (1996) propose a model of innova-tion implementation success explained by both the formal mechanisms creating a climate for the innovation's implementation (practical compatibil-ity) and the fit of an innovation to the targeted users' values at the organiza-tional and group level (value compatibility). Few, if any, researchers have investigated the dual nature of compatibility, despite its promising potential contribution to our understanding of successful implementation of new technologies (Tornatzky & Klein, 1982).

Yet some findings suggest a critical factor in the success of telecommuting arrangements is both "a supportive culture and appropriate systems" (Baruch & Nicholson, 1997, p. 26). Similarly, Suomi, Luukinen, Pekkola and Zamindar (1998) state that "adequate technological and communication infrastructure, sufficient technological skills and a reorganization of corporate structures and cultures is needed to provide for successful and lasting telework arrangements" (p. 355).

Therefore, a primary objective of this study is to understand further whether the reason for the slow growth of telecommuting is due to a lack of the organizations' values compatibility, practical compatibility, or both. If managers see telecommuting as incompatible with their existing values, employee skills, and work practices, telecommuting may never reach the previous claims for its success.

PRACTICAL COMPATIBILITY

IS research has found support for a positive influence of practical compatibility, such as IS sophistication and IS resources on Information Technology (IT) adoption, use, and success (Premkumar & Ramamurthy, 1995; Tornatzky & Klein, 1982). Moreover, the information architecture, technology infrastructure, and IS human resources are of primary concern to top MIS managers (Niederman, Brancheau, & Wetherbe, 1991). Successful telecommuting is believed to be related to the ability of employees to telecommute, the existing hardware and software, and managerial competence (Belanger & Collins, 1998; Nilles, 1997).

Employees' ability to telecommute is important, not only because telecommuting employees must work on their own without access to technology experts, but also because they need to interact effectively with each other and their managers when face-to-face contact is reduced (Belanger & Collins, 1998; Nilles, 1997). Performance criteria must be worked out, and a belief in employees as competent and reliable must exist; otherwise, managers' concerns about how to measure telecommuters' performance may act as a barrier to telecommuting (Christensen, 1992; Guthrie, 1997; Nilles, 1997).

While there is no specific list of technologies required for successful telecommuting, appropriate IT tools (such as e-mail, electronic white boards, and voice mail) act to increase the effectiveness, not only of individual telecommuters, but also of their fellow team members (Nilles, 1997). Hardware and software needs must also address system security. The fear of intrusion from outsiders, the potential storage of data off-site, and inadvertent leakage from employees may be some of the early barriers to the adoption of telecommuting (Belanger & Collins, 1998; DiMartino & Wirth, 1990; Nilles, 1997).

While some evidence suggests the need for increased security, others consider it to be a major disadvantage of telecommuting (cf. DiMarino & Wirth, 1990; Nilles, 1997; Ruppel & Howard, 1998). Moreover, many managers believe that external networking poses a major organizational risk; one study (i.e., Loch, Carr, & Warkentin, 1992) found that many managers do not believe external networking poses a major threat because the point of entry usually is a well-secured mainframe. Thus, those who feel their security measures are sufficient, such as those with mainframes, would be more likely to believe telecommuting is compatible with their organization and would be more likely to implement it. Thus,

H1: The greater the practical compatibility of telecommuting with the IS organization, the greater the level of adoption, diffusion and success of telecommuting.

VALUE COMPATIBILITY

Management approaches to distributed work arrangements may be determined by the arrangements' compatibility with the organization's values (Belanger & Collins, 1998). These values include attitudes toward control mechanisms that maximize performance and manifest themselves in activities such as monitoring employee behaviors, rewarding employees for productivity, encouraging group behaviors and outcomes, or enabling employee autonomy (Belanger & Collins, 1998). A major concern in distributed arrangements has been whether employees are sufficiently empowered while having specific goals (Belanger & Collins, 1998).

Quinn and Rohrbaugh (1981) developed a framework for organizational effectiveness based on competing organizational values. This framework has been used to study organizational cultures and their associated innovations (e.g., Denison & Spreitzer, 1991; Yeung, Brockbank & Ulrich, 1991; Zammuto & Krakower, 1991) because of its ability to tap into the aspects of organizational effectiveness via different values, assumptions, and interpretations that define an organization's culture. The framework suggests that although an organization may contain multiple values, one cultural set of values generally prevails.

Organizational values have been found to influence the successful adoption of several IT innovations, including CASE, Lotus Notes, and advanced manufacturing technologies (Orlikowski, 1993a, b; Zammuto & O'Connor, 1992). Yet Cooper (1994) suggests that organizational culture and its associated values comprise an area that largely has been ignored by IT implementation researchers. Romm, Pliskin, Weber, and Lee (1991) suggest that without a match between the values of an organization and the value assumptions embedded in an IT innovation, a costly implementation failure will likely occur.

The set of value dimensions described by Quinn and Rohrbaugh's (1981) competing values framework are (1) people vs. the organization, (2) stability and control vs. change and flexibility, and (3) means vs. ends. Based on these dimensions, four value orientations were identified: (1) *group*: people, flexibility, and human resource development; (2) *developmental*: organization, flexibility, growth, and inventiveness; (3) *rational*: organization, control, goal setting, outcomes, competence, and efficiency; and (4) *hierarchical*: people, control, conservative, procedures, and rule oriented (cf. Zammuto & Krakower, 1991).

Group Values

It would be expected that group values would be conducive to telecommuting. Group values emphasize human resources and member participation in decision making and flexibility; both are considered important to successful telecommuting (Nilles, 1994). Management openness to ideas and suggestions in decision making leads to higher levels of loyalty, trust, and long-term commitment (Graen & Cashman, 1975; Whitener et al., 1998; Zammuto & Krakower, 1991). Organizations that have the values of sharing as well as attention to human relationships and employee interests are more likely to achieve open communication among employees.

Such values encourage dialogue that reinforce shared norms of protecting mutual interests and that lead to commitment in implementing well-understood solutions (Schein, 1996). One telecommuting study found that "pre-planning of the communication process may be a necessary precondition for the successful implementation of telework arrangements" (Duxbury & Neufeld, 1999, p. 24). Therefore, values like group values that already emphasize communication would be more compatible with telecommuting arrangements. Moreover, relations between managers and telecommuters may deteriorate without continued feedback sessions, team building activities, and other forms of communication (Reinsch, 1997).

Bernardino (1997) found that most employers are motivated to offer telecommuting to address employee needs, thereby suggesting attention to employee interests and group values. Thus, group values emphasize communication, teamwork, and flexibility that would encourage the implementation of telecommuting. Therefore,

H2: The stronger the group values of the organization, the greater the level of adoption, diffusion and success of telecommuting.

Developmental Values

Developmental values are characterized by environmental scanning and an assumption of change, but emphasize the organization over the individual. The importance or ideological appeal of the task being undertaken motivates individuals (Zammuto & Krakower, 1991). As previously discussed, telecommuting has many potential benefits for the organization, such as increased productivity, office space savings, and other cost savings. Nilles (1997) suggests that telecommuting's growth has been limited to date because it is one of the better-kept secrets in business and that "high benefits with comparatively low cost and risk mean a don't-tell policy regarding competitors" (Nilles, 1997, p. 14).

Bernardino (1997) found that a familiarity with the telecommuting option increases the likelihood to 94.1% that a telecommuting program is offered, and it also suggests that flexible and ad hoc organizations and arrangements are particularly appropriate for telecommuting. Therefore, developmental values emphasizing flexibility and inventiveness would be conducive to the adoption of telecommuting out of a sense of competition.

H3: The stronger the developmental values of the organization, the greater the level of adoption, diffusion, and success of telecommuting.

Rational Values

Rational values reflect an underlying belief in the need to perform analytical appraisals of performance following clear statements of purpose and targets. These values emphasize goals and management evaluation of performance (Greenwood & Hinings, 1993). Where outcomes-based measures are used, telecommuting success is likely to be enhanced, for employee performance monitoring no longer is based on the employee being present or simply appearing busy (Belanger & Collins, 1998; DiMartino & Wirth, 1990). Similarly, where the values emphasize output rather than process, greater delegation and trust have occurred and resulted in higher levels of performance (El Sawy, 1985; Jarvenpaa, et al., 1998). Moreover, rational values, similar to developmental values, would be more susceptible to arguments emphasizing the efficiency and objective side of telecommuting, particularly in the adoption stage (Cooper & Zmud, 1990). Therefore,

H4: The stronger the rational values of the organization, the greater the level of success of telecommuting.

Hierarchical Values

An innovation may fit within an existing culture or it may be countercultural. Telecommuting may be countercultural in organizations that exhibit hierarchical values, which may exist, in part, to control personnel. Both rational and hierarchical values focus on control, but rational values emphasize outcomes and goals, whereas hierarchical values emphasize procedures and rules. Clearly, focusing on outcomes and goals is more congruent with telecommuting, and managers in organizations with such values should adapt more easily to the idea that employees will no longer be monitored on a day-to-day basis.

Researchers (e.g., El Sawy, 1985; Sitkin & Roth, 1993) suggest that organizations frequently adopt formal rules when trust is lacking, but such remedies are ineffective in building trust because they do not address value differences. Handy (1995) suggests that to obtain the efficiencies and other benefits of virtual arrangements, organizations must be based on trust rather than control mechanisms. Because of the hierarchical control, large organizations are not conducive to trust, and it is better to devise smaller, fairly constant employee groupings in such organizations (Handy, 1995). Similarly, Nilles (1997) states that traditional offices tend to be best suited to hierarchical structures. Therefore,

H5: The stronger the hierarchical values of the organization, the lesser the level of adoption, diffusion and success of telecommuting.

METHODOLOGY

Sample

IS personnel, with their IT-related knowledge and relative independence in completing their tasks, are believed to be well suited to telecommuting (Baruch & Nicholson, 1997). Therefore, surveys were sent to IS managers in 1,200 different organizations. Approximately 900 surveys reached the IS manager, as estimated by a reminder, follow-up survey with address correction requested. The survey resulted in 125 usable responses, for a response rate of approximately 14 percent.

It was believed that these IS managers would be in the best position to observe both telecommuting's practical and value compatibility with their organizations and the level of telecommuting by IS personnel. Also, any policies concerning telecommuting adoption and diffusion are likely to be under the IS manager's control, and the IS manager may be in the best position to determine the overall success of telecommuting.

To confirm or refute this assumption, the respondents were asked, "What proportion of the decisions about the MIS department's use of new technologies, such as ... telecommuting, are you responsible for?" The average of the estimates was approximately 81 percent of the decisions, with nearly half of the respondents indicating they had been in the position to make such decisions for between five and 10 years. Therefore, the questionnaire appears appropriately targeted.

In addition, the respondents represented companies at a variety of levels: local level (20%), national level (40%), and international level (40%). A mix of mainframe and PC users was represented with approximately 18% of the respondents mainframe-oriented, 15% PC-oriented, and 67% both mainframe and PC-oriented.

To determine if a representative sample was achieved, two tests for non-response bias were conducted. First, a follow-up survey was sent asking non-respondents the reason they did not respond. The major reasons given for not responding were as follows: (1) too many surveys (36%), (2) not enough time (23%), and (3) length of survey (20%). Second, industry classifications of the respondents were compared to the industry classification of the mailing list; no significant departures were found. Also, the survey asked about technologies other than telecommuting, so the sample is not solely biased towards those organizations that telecommute. Therefore, the researchers do not believe there is a non-response bias, even though the response rate is less than desired.

Sixty-nine (69) respondents reported some degree of telecommuting, and 30 reported some knowledge of telecommuting results. The surveys were analyzed using correlation analysis, followed by multivariate regression to test the level of variance in telecommuting explained by the independent variables.

Measures

Practical compatibility. Compatibility was measured by respondents' ratings of telecommuting's fit with their organizations' resources on a scale of 1 = very poor to 5 = very good. The phrases used were "fit with hardware and software currently in use," "ability to control or secure telecommuting," and "ability of MIS personnel to use or implement." While the three items loaded on one factor in a factor analysis, it was apparent that the "ability to control or secure telecommuting" was negatively impacting the Cronbach alpha. Therefore, the "ability to control or secure telecommuting" was separated from the practical compatibility variable and treated separately through the rest of the analysis. The resulting Cronbach alpha for the remaining two items is 0.83.

Values compatibility. Measures for value orientation were taken from Yeung, Brockbank and Ulrich's (1991) competing values instrument. Respondents were asked to rate, on a scale of 1 = very strongly disagree to 7 = very strongly agree, the extent to which each statement described their IS department. Cronbach reliabilities for the four value types ranged from 0.66 to 0.68. Examples of the items for each values type follow:

- Group: "My MIS department is a very *personal* place. It is like an extended family. People seem to share a lot of themselves" (3 items).
- Developmental: "My MIS department is a very *dynamic and entrepreneurial* place. People are willing to stick their necks out and take risks" (3 items).
- Rational: "My MIS department is a very *production oriented* place. People are concerned with getting the job done" (3 items).
- Hierarchical: "My MIS department is a very *formal and structured* place. People pay attention to procedures to get things done" (2 items).

Telecommuting adoption, diffusion, and success. The measure for the adoption of telecommuting was a four-point scale: 1 = "considered at one time and rejected," 2 = "currently being considered for use but not in use," 3 = "currently being used on a trial basis," and 4 = "implemented." Those who marked an item which read "not familiar with the practice or never considered it" were eliminated from further analysis since they would not have considered the practical compatibility associated with telecommuting.

Diffusion was a six-point scale: -1 = "tried and rejected," 0 = "considered for use, but not in use yet," 1 = "initial or sporadic use," 2 = "a few people use regularly; slightly implemented," 3 = "many people use regularly; partially implemented," and 4 = "all people use regularly; fully implemented." Telecommuting success was measured on a four-point scale with results ranging from poor to excellent. The measurement of telecommuting success depended on the respondent having already adopted telecommuting. It should be noted that since the measure of success is perceptual, the researchers validated it by correlating it with a more objective ROI measure. This relationship was significant at the .05 level, thus suggesting that the perceptual measure, often used in innovation research, is capturing the success construct.

RESULTS

Correlation results (Table 1) clearly show a relationship between the ability to secure telecommuting and adoption and diffusion of telecommuting. Group values are related to telecommuting diffusion and results.

Seventy percent of mainframe managers, 47% of PC managers, and 69% of managers of both mainframes and PCs felt their organization had a poor ability to secure or to control telecommuting. This result is opposite of that found by Loch et al. (1992), where mainframe managers felt their security procedures were well suited to external access.

Stepwise regression analysis (Table 2) was used to examine the explanatory power of the variables found to be significant in the correlation analysis. Variables significant in adoption included the ability to secure telecommuting and developmental values, with 24% of the variation explained. Group values, the ability to secure telecommuting, and developmental values were significant variables in diffusion, with 49% of the variation explained. Practical compatibility (without security), group culture, and rational values explained 52% of the variation in telecommuting results.

Table 1: Spearman correlation coefficients

	1	2	3	4	5	6	7	8
1. Ability to secure								
2. Practical compatibility	0.28*							
3. Developmental values	-0.12	0.23[+]						
4. Group values	0.12	0.11	0.19[+]					
5. Rational values	0.08	0.11	0.42***	0.37***				
6. Hierarchical values	0.20	-0.21	0.03	-0.05	0.03			
7. Adoption	0.39**	0.07	0.20	0.20	0.20	-0.10		
8. Diffusion	0.44**	0.16	0.25	0.57***	0.30[+]	0.09	0.87***	
9. Results	0.27	0.21	0.00	0.50**	-0.01	-0.11	0.36*	0.65***

[+]p<.10, *p<.05, **p<.01, ***p<.001

Table 2: Stepwise regressions

Dependent/Independent Variable	F Value	P Value*	Partial R-Squared	Model R-Squared	Model F-Value	Pr>F
Telecommuting Adoption				0.24	7.49	0.0015
Ability to secure	11.47	0.001	0.19			
Developmental values	3.04	0.09	0.05			
Telecommuting Diffusion				0.49	10.79	0.0001
Group values	14.47	0.001	0.29			
Ability to secure	9.88	0.004	0.16			
Developmental values	2.84	0.10	0.04			
Telecommuting Results				0.52	8.69	0.0004
Practical compatibility (without ability to secure)	9.84	0.004	0.28			
Group values	8.47	0.007	0.18			
Rational values	3.14	0.09	0.06			

*P value to enter was left at SAS's default value of p=0.15

LIMITATIONS OF THE STUDY

The study was cross-sectional and retrospective, and so causation was not confirmed. Nonetheless, an attempt was made to study three states of telecommuting: adoption, diffusion, and results, which had the effect of studying organizations in various states of telecommuting implementation in a pseudo-longitudinal form.

CONCLUSIONS

This study is an attempt to see whether an organization's practical compatibility and value compatibility with telecommuting influence each of the three dependent variables of adoption, diffusion or success of telecommuting. The findings suggest the types of compatibility and their dimensions influence the dependent variables differently.

The ability to secure telecommuting arrangements (a dimension of practical compatibility) strongly affects both the adoption and diffusion of telecommuting. The strength of these correlations suggests that managers are looking at the practical side of telecommuting and are more likely to adopt it if they believe their organization can ensure security. Despite previous research that mainframe managers are more confident of their existing security systems, a higher proportion of mainframe managers in this study believed their systems were poor in ability to secure telecommuting. This finding may be the result of a greater awareness of security issues since earlier studies, or it may result from the fact that telecommuting is an application that does not fit the traditional nature of mainframe security. Although mainframe vs. PC security is not a focus of the study, further investigation is needed to understand this unexpected result and the impact and nature of security needs for telecommuting. In addition, further investigation into the organization's ability to secure the IT as one form of practical compatibility takes on added importance in IS research as other similar virtual arrangements occur.

While security concerns pose a barrier to both the initial adoption and the diffusion of telecommuting, values compatibility takes on increasing importance during the telecommuting diffusion process. Group values, in particular, become increasingly important and may indicate that, as employees work remotely with less contact time, keeping communication going and group processes take on added importance. These results are consistent with Reinsch (1997), who found that organizational neglect of the telecommuter-manager relationship may become more apparent as telecommuting becomes more widespread and as increasing numbers of telecommuters lose social support and coaching from coworkers and managers. Managers of telecommuters must be flexible enough to adapt to managing from a distance, which generally involves a significant change from managing in the traditional method (Belanger & Collins, 1998; Nilles, 1997). Thus, managers who supervise telecommuters successfully must be able to adapt communication methods to those most appropriate when physical distance exists (e.g., e-mail, fax and telephone) (Duxbury & Neufeld, 1999). The implication is that managers must provide encouragement and ways for employees to continue communicating after the initial adoption. Also, managers who wish to realize the best possible results from telecommuting should ensure that the human resource practices and group values are in place.

The significant role of group values may also help explain why previous alternative work arrangement or telecommuting studies (e.g., Kavan & Saunders, 1998; Ruppel & Harrington, 1995) found that middle manager resistance acts as a primary barrier to adoption. Even in situations where telecommuting provides benefits, such as increased employee productivity, increased employee morale, lower turnover and absenteeism, and reduced office and parking space needs, which accrue to the organization and the manager, managers resist adoption (Kavan & Saunders, 1998). Based on the findings of their 1998 study, Kavan and Saunders proposed that managers' resistance needed more research and may be due to a lack of a compatible corporate culture or of alternative work arrangements. When managers were allowed to adapt their human resource practices to be more equitable and group-oriented, alternative work arrangements were more likely to be adopted by managers. Where group values are already a strong part or are becoming a part of the corporate culture and control mechanism, management resistance may be reduced, even if it involves significant change/adaptation on the part of the managers. Thus, the current study provides further clarification of Kavan and Saunders' results. Those organizations that value people, flexibility, open participation, and human resource development have values more compatible with telecommuting arrangements, and these values likely lead to less management resistance.

There is some support, albeit weak, for developmental values as compatible with adoption and diffusion and rational values as compatible with telecommuting diffusion and results. Developmental values, which emphasize flexibility and change, would likely help middle managers adapt to the change embedded in telecommuting. Rational values would play a role in the diffusion and results, presumably because organizations that manage by setting goals, rather than by monitoring work effort, would be better able to integrate telecommuters into their existing management style. Management's perceived loss of control, reported to be an obstacle to telecommuting (i.e., Handy, 1997, 1998), may be less of an obstacle in organizations with rational values. Further research, however, is needed to verify the existence of the weak effects found for these two organizational value sets.

The relationship between telecommuting success and practical compatibility, with its dimensions of ability of IS personnel to use or implement telecommuting and telecommuting's fit with the hardware and software currently in use, suggests that training and IT infrastructure may be important to telecommuting's success. While this form of practical compatibility does not appear to pose a barrier to telecommuting adoption or diffusion, it may be important in achieving the maximum return from telecommuting. This is consistent with Nilles' (1997) suggestion that training and existing information technology play an important role in achieving success. Moreover, a greater investment in technology may facilitate communication and activities in line with group values, which also were found to be a facilitator of telecommuting diffusion and success in this study.

The findings resulting from segregating practical compatibility from value compatibility and further dividing practical compatibility suggest that the compatibility of telecommuting with an organizational environment is a very complex issue. This complexity makes conducting research in telecommuting, as well as explaining its slow growth, a difficult task that requires further well-defined research. As an extension of this research, the many aspects of compatibility must be considered and matched with the type of telecommuting arrangement because not all types of arrangements are equally compatible with an organizational environment. For example, Bernardino (1997) studied the different types of telecommuting arrangements (i.e., number of days per week, flexible vs. fixed hours) and found that the differences in arrangements are affected by the organization's characteristics, such as structure. The matching of various organizational characteristics with various telecommuting arrangements to determine their compatibility is complex and a fruitful area for further research.

This study has implications for those wishing to champion telecommuting. Generally, the fit between the organization's values and practices must be examined to determine the path of least resistance and to assist in implementation success. Those wishing to champion telecommuting in their organizations need to take practical steps, such as making sure the telecommuting arrangement can be secured. Group values should be advocated to aid in telecommuting's diffusion and success, and the investment in people, hardware, and software to support the use of the technology should be adequate.

Therefore, this study shows that IS studies of IT innovations could benefit greatly from distinguishing between practical and values compatibility and between adoption, diffusion, and success of the IT innovation. While practical compatibility appears to influence adoption and diffusion, values compatibility, especially those related to the human resource practices of the organization, increasingly aid diffusion and results. The possibility that the different role of these two forms of compatibility influences other IT tools in a similar manner should provide a fruitful area for future research.

IMPLICATIONS FOR MANAGEMENT

Telecommuting has potential benefits for all parties to the relationship if it is designed in a way that maximizes the benefits while minimizing the potential pitfalls. One of these potential pitfalls is a perceived loss of management control of employees. While this perception may not match reality, it must be minimized to allow the arrangement to flourish and be viewed as successful.

Managers who do not "fear" telecommuting arrangements are, where appropriate, more likely to encourage employees to use this option and jointly work toward its success. This study suggests several ways managers can become less "fearful" and can increase the compatibility, and thus success, of telecommuting arrangements in their organizations.

Some of the compatibility issues studied here are easier, or at least quicker, to implement than others. Generally, those issues involving practical compatibility will be easier to accomplish in the short run than those involving value compatibility. If the organization does not already have telecommuting arrangements (has not adopted them), then it should examine and, if appropriate, increase security measures. From a values perspective, managers should be encouraging adaptability to change among organizational members. Guidelines or ideas for change may come from actively scanning the environment, particularly competitors who may be using telecommuting arrangements and receiving the benefits potentially available.

If telecommuting arrangements have been adopted but are not being widely used effectively throughout the organization, possible areas to scrutinize are the ability to secure these arrangements and human resources, or group values. An emphasis on employee needs, empowerment of employees, a trusting attitude towards employees and open lines of communication are important. These types of organizational values are more difficult to develop in the short term; however, management can send signals designed to build these values. For example, management can initiate frequent, honest communication, employee empowerment, and team-based problem-solving approaches.

Showing that telecommuting arrangements can be or are successful is also important. Such success is related to rational values and practical compatibility, which are concerned with the use of objective productivity measures and the ability of the employees to use the technology. In addition to this study, much of the popular telecommuting literature suggests that successful telecommuting arrangements are related to well-planned, formal agreements that clearly outline responsibilities on the part of both the manager and the employee. Thus, success can be accomplished through formal arrangements with clear metrics, by providing training to managers in how to manage from a distance, and by providing employees with training on equipment, systems, communicating, and time management strategies.

Therefore, managers must adopt strategies over time to aid in the adoption, use, and perceived success of telecommuting arrangements. The ideas presented here are not incompatible with one another; thus, managers can begin initiatives to stress these goals as the process proceeds. For example, securing systems to allow telecommuting is important in both the adoption and use of telecommuting and should be a primary concern. Also, to aid in the adoption, the potential advantages of telecommuting need to be stressed as well as the competitive reasons for doing so. In anticipation of building a successful program, formal arrangements can be devised and training can be conducted. This approach will also have the effect of increasing communication and beginning to build the human resources/group values that will allow telecommuting arrangements to achieve their potential.

REFERENCES

Baruch, Y., & Nicholson N. (1997). Home, Sweet Work: Requirements for Effective Home-working. *Journal of General Management,* 23(2), 15-30.

Belanger, F., & Collins, R. W. (1998). Distributed Work Arrangements: A Research Framework. *The Information Society,* 14, 137-152.

Bernardino, A. (1996). *Telecommuting: Modeling the Employer's and the Employee's Decision-Making Process.* New York & London: Garland Publishing.

Bresnahan, J. (1998). Why Telework? *CIO,* 11(7), 28-37.

Christensen, K. (1992). Managing Invisible Employees: How to Meet the Telecommuting Challenge. *Employee Relations Today,* 19(2), 133-143.

Cooper, R. (1994). The Inertial Impact of Culture on IT Implementation. *Information & Management,* 27(1), 17-31.

Cooper, R. B., & Zmud, R. W. (1990). Information Technology Implementation Research: A Technological Diffusion Approach. *Management Science,* 36(2), 123-139.

Denison, D., & Spreitzer, G. (1991). Organizational Culture and Organization Development: A Competing Values Approach. In R. W. Woodman & W. A. Pasmore (Eds.), *Research in Organizational Change and Development* (Vol. 5, pp. 1-21). Greenwich, CT: JAI Press Inc.

DiMartino, V., & Wirth, L. (1990). Telework: A New Way of Working and Living. *International Labour Review,* 129(5), 529-554.

Downs, G. W., & Mohr, L. B. (1976). Conceptual Issues in the Study of Innovation. *Administrative Science Quarterly,* 21(4), 700-714.

Duxbury, L., & Neufeld, D. (1999). An Empirical Evaluation of the Impacts of Telecommuting on Intra-organizational Communication. *Journal of Engineering and Technology Management,* 16(1), 1-28.

El Sawy, O. A. (1985). Implementation by Cultural Infusion: An Approach for Managing the Introduction of Information Technologies. *MIS Quarterly,* 9(2), 131-140.

Graen, G., & Cashman, J. F. (1975) A Role-Making Model of Leadership in Formal Organizations: A Developmental Approach. In J. Hunt & L.

Larson (Eds.), *Leadership Frontiers* (pp. 143-165). Kent, OH: KSU Press.

Greenwood, R., & Hinings, C. R. (1993). Understanding Strategic Change: The Contribution of Archetypes. *Academy of Management Journal,* 36(5), 1052-1081.

Guthrie, R. (1997). The Ethics of Telework. *Information Systems Management,* 14(4), 29-32.

Handy, C. (1998, March). Truths That Are Hard to Live With. *Management Today,* 31.

Handy, C. (1997). Unimagined Futures in Hesselbein. In M. Goldsmith & R. Bechard (Eds.), *The Organization of the Future* (pp. 377-383). San Francisco: Jossey-Bass Publishers.

Handy, C. (1995). Trust and the Virtual Organization. *Harvard Business Review,* 73(3), 40-50.

Jarvenpaa, S. L., Knoll, K., & Leidner, D. E. (1996). Is Anybody Out There? Antecedents of Trust in Global Virtual Teams. *Journal of Management Information Systems,* 14(4), 29-64.

Jones, D. (1996, November 25). Telecommuting Honks Own Horn: Few Hop Aboard. *USA Today,* p. 1B.

Kavan, C. B., & Saunders, C. S. (1998). Managers: A Key Ingredient to Alternative Work Arrangements Program Success. *Journal of End User Computing,* 10(4), 23-32.

Klein, K. J., & Sorra, J. S. (1996). The Challenge of Innovation Implementation. *The Academy of Management Review,* 21(4), 1055-1080.

Loch, K. D., Carr, H. H., & Warkentin, M. E. (1992). Threats to Information Systems: Today's Reality, Yesterday's Understanding. *MIS Quarterly,* 16(2), 173-186.

Niederman, F., Brancheau, J. C., & Wetherbe, J. C. (1991). Information Systems Management Issues for the 1990s. *MIS Quarterly,* 15(4), 475-495.

Nilles, J. M. (1997). Telework: Enabling Distributed Organizations. *Information Systems Management,* 14(4), 7-14.

Nilles, J. M. (1994). *Making Telecommuting Happen: A Guide for Telemanagers and Telecommuters.* New York: VanNostrand Reinhold.

Orlikowski, W. J. (1993a). Learning from Notes: Organizational Issues in Groupware Implementation. *Information Society,* 9(3), 237-251.

Orlikowski, W. J. (1993b). CASE Tools as Organizational Change: Investigating Incremental and Radical Changes in Systems Development. *MIS Quarterly,* 17(3), 309-340.

Premkumar, G., & Ramamurthy, K. (1995). The Role of Interorganizational and Organizational Factors on the Decision Mode for Adoption of Interorganizational Systems. *Decision Sciences,* 26(3), 303-336.

Quinn, R. E., & Rohrbaugh, J. (1981). A Competing Values Approach to Organizational Effectiveness. *Public Productivity Review,* 5(2), 122-140.

Quinn, R. E., & Rohrbaugh, J. (1983). A Spatial Model of Effectiveness Criteria: Towards a Competing Values Approach to Organizational Analysis. *Management Science,* 29(3), 363-377.

Reinsch, N. L. (1997). Relationships Between Telecommuting Workers and their Managers: An Exploratory Study. *Journal of Business Communication,* 34(4), 343-369.

Rogers, E. M. (1983). *Diffusion of Innovations.* New York: Free Press.

Romm, C. T., Pliskin, N., Weber, Y., & Lee, A. (1991). Identifying Organizational Culture Clash in MIS Implementation: When is it Worth the Effort? *Information & Management,* 21(2), 99-109.

Ruppel, C. P., & Harrington, S. J. (1995). Telework: An Innovation Where Nobody is Getting on the Bandwagon? *Data Base for Advances in Information Systems,* 26(2 & 3), 87-104.

Ruppel, C. P., & Howard, G. S. (1998). Facilitating Innovation Adoption and Diffusion: The Case for Telework. *Information Resources Management Journal,* 11(3), 5-15.

Schein, E. H. (1996). Three Cultures of Management: The Key to Organizational Learning. *Sloan Management Review,* 38(1), 9-20.

Sitkin, S. B., & Roth, N. L. (1993). Explaining the Limited Effectiveness of Legalistic 'Remedies' for Trust/Distrust. *Organization Science,* 4(3), 367-392.

Soumi, R., Luukinem, A., Pekkola, J., & Zamindar, M. (1998). Telework — The Critical Management Dimension. In P. J. Jackson & J. M. van der Wielen (Eds.), *Teleworking: International Perspectives From Telecommuting to the Virtual Organisation* (pp. 329-336). London & New York: Routledge.

Tornatzky, L. G., & Klein, K. J. (1982). Innovation Characteristics and Innovation Adoption-Implementation: A Meta-Analysis of Findings. *IEEE Transactions on Engineering Management,* EM-29(1), 28-45.

Whitener, E. M., Brodt, S. E., Korsgaard, M. A., & Werner, J. M. (1998). Managers as Initiators of Trust: An Exchange Relationship Framework for Understanding Managerial Trustworthy Behavior. *Academy of Management Journal,* 23(3), 513-530.

Yeung, A. K. O., Brockbank, J. W., & Ulrich, D. O. (1991). Organizational Culture and Human Resource Practices: An Empirical Assessment. In R. W. Woodman & W. A. Pasmore (Eds.), *Research in Organizational Change and Development* (Vol. 5, pp. 59-81). Greenwich, CT: JAI Press Inc.

Zammuto, R. F., & Krakower, J. Y. (1991). Quantitative and Qualitative Studies of Organizational Culture. In R. W. Woodman & W. A. Pasmore (Eds.), *Research in Organizational Change and Development* (Vol. 5, pp. 83-114). Greenwich, CT: JAI Press Inc.

Zammuto, R. F., & O'Connor, E. J. (1992). Gaining Advanced Manufacturing Technologies' Benefits: The Roles of Organization Design and Culture. *Academy of Management Review,* 17(4), 701-728.

ENDNOTE

1   The order of authors was randomly determined. Each author contributed equally to this paper. Funding was provided through the Faculty Research Award program of the Georgia College & State University Graduate School and Research Services department and through the Academic Challenge program at The University of Toledo.

2   Telecommuting is defined as the employees' "use of telecommunications equipment to carry out their normal day-to-day activities while physically located off site from the standard workplace."

ACKNOWLEDGEMENT

The authors are grateful to Lei Jin for her invaluable assistance on this project.

Note: An earlier version of this research was presented at ICIS 1999.

Chapter VII

Telecommuter Selection: A Systems Perspective

Janet A. Henquinet
Metropolitan State University, USA

ABSTRACT

This chapter presents a conceptual framework for the telecommuter selection process. The framework uses a systems perspective to identify critical variables and relationships in selecting effective telecommuters. The purpose of the model is twofold: to assist managers and organizations in developing selection procedures and to identify opportunities for future research.

INTRODUCTION

Organizations experiment with telecommuting to save costs, increase employee satisfaction, improve productivity, tap nontraditional labor pools, and for many other reasons (Apgar, 1998; DiMartino, 1990; Greengard, 1994; Kugelmass, 1995; Piskurich, 1996). In a 1997 survey by Robert Half International, 35 percent of the respondents predicted there would be a strong increase in the number of telecommuting employees in their organization and 52 percent indicated some increase in telecommuting (Messmer, 1998). In the United States, telecommuting is becoming more common in the private sector and federal, state and local government units are implementing telecommuting programs at a significant rate (Mahfood, 1994; McCune, 1998; Pynes, 1997). Given the predicted growth of telecommuting, effective telecommuter selection will become more important in the future.

While telecommuting in the broadest definition includes workers in any remote location, much of the writing focuses on employees who perform jobs from home (Mahfood, 1994). The term telecommuting is often synonymous with terms such as teleworking and homeworking. The three terms will be used interchangeably and the focus will be on individuals who are employed by an organization but work at home on a full- or part-time basis. This is a distinctly different context than mobile workers who may use client offices, remote/satellite work centers, or their cars and hotel rooms for work tasks (Chapman, Sheehy, Heywood, Dooley, & Collins, 1995; Greengard, 1994; Kugelmass, 1995).

This chapter provides a conceptual framework for telecommuter selection. A systems perspective identifies critical variables and relationships to help develop better selection procedures and to suggest opportunities for research.

A SYSTEMS PERSPECTIVE

A review of the current literature indicates the advantages of taking a systems perspective of the selection process. This approach emphasizes the interactive and dynamic aspects of organizational activities as they relate to telecommuting (Chapman et al., 1995). Schuler and Jackson (1996, p. 259) note that selection "must be congruent with the internal and external environment, integrated with other human resource activities, and done in a manner consistent with legal regulations."

Baruch and Nicholson (1997), in a study of homeworking in the UK, stress the need to focus on four realms: the job, the individual, the home/work interface and the organization. Moorcroft and Bennett (1995) succinctly state the need for the right people, the right managers, and the right jobs.

Successful performance is determined in great part by the work context. For the telecommuter selection process to be effective, it is important to look not only at employee characteristics but also to examine job characteristics, supervisor characteristics, and organizational support systems. Figure 1 illustrates how these four elements can be aligned to ensure a comprehensive and effective selection process.

Employee Characteristics

Desirable employee characteristics are often the first topic in telecommuter selection discussions and that results in a profile of the ideal telecommuter.

Figure 1: A systems framework for selection of telecommuters

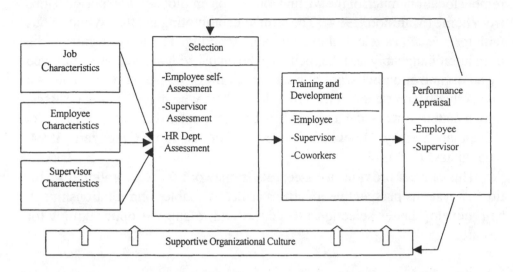

Only a few empirical studies have addressed this issue (Baruch, 1999; Belanger, 1999), but the practitioner literature provides anecdotal evidence from organizations and prescriptive advice from writers in the field. Employee characteristics emphasized are personality traits, employment status, and family/home situation.

Personality traits. "Telecommuters are made not born. There are certain character traits, however, that lend themselves to making a successful telecommuter" (Mahfood, 1994, p. 31). Commonly cited personal characteristics of the ideal telecommuter include self-motivation, self-discipline, results orientation, and ability to work independently (Hamilton, 1987). Chapman et al. (1995) indicate that selection criteria for teleworkers usually are similar to those for other workers, with the one exception being more focus on self-management skills: capacity to organize work schedules, establish priorities, meet deadlines and self-assess performance. Mahfood (1994) stresses the importance of effective communication skills. Reilly (1997) ranks motivation, self-guidance, and technological literacy as important factors. A higher level of technological literacy may be necessary due to lack of easily available technical support when working at a remote location.

Baruch and Nicholson (1997) asked homeworkers in the UK to identify personal qualities needed to be a successful homeworker. The respondents cited self-motivation (45%), ability to work alone (42%), tenacity (29%), and being organized (29%) as most important. Two factors were detrimental: need for social life (37%) and need for supervision (35%). Other problems include feelings of isolation, distractions (television, family, and refrigerator) and difficulty separating home and work life (Bernardi, 1998).

Employment status. Employment status is another frequently used selection criterion. Pynes (1997) recommends that only employees with experience in the particular job and with a satisfactory or higher performance rating be considered for telecommuting. Experience with the company may help to ensure that employees have developed commitment to the organization (Hamilton, 1987). For example, to be considered for telecommuting at AEGON, an employee must have one year of company service, six months in the current position, an above average performance rating, a satisfactory attendance record, and the manager's approval (McGonegle, 1996).

Family and home situation. Telecommuter applicants often are screened for an appropriate home situation (McGonegle, 1996; Mahfood, 1994). AEGON's process, mentioned previously, identified three critical issues: work status of the spouse, presence of children/dependents, and availability of workspace. It requires employees to have adequate space and to have dependent care arrangements (McGonegle, 1996). Baruch and Nicholson (1997), in their interviews of homeworkers, noted a decrease in stress associated with work, but family-related stress increased significantly. Hamilton (1987) suggests that family members be interviewed. Mahfood (1994) stresses the need to educate family and friends to recognize and support the new rules that come with telecommuting. IBM's virtual office training provides guidelines and specific techniques for family management (Davenport & Pearlson, 1998). The need for empirical study is highlighted by Hartman, Stoner and Arora (1991) who found that family disruption did not significantly impact telecommuting productivity but there was a significant negative correlation between family disruption and telecommuter satisfaction.

Privacy concerns often arise when discussing the home and family context (European Foundation, 1997; Sudbury & Towns, 1997). This is a familiar issue for international organizations that expatriate employees (move an employee and family to another country for a work assignment). Successful selection includes assessing the family situation within the context of legal and ethical considerations (Briscoe, 1995; Dowling, Schuler, & Welch, 1994). Two of the top three reasons for expatriate failure are related to the family (Tung, as cited in Briscoe, 1995). Study of the expatriate model may inform the telecommuter selection process.

Kugelmass (1995) warns about an inappropriate emphasis on the telecommuter profile: "The emphasis on choosing the right people, of making highly individualized assessments of an employee's suitability, and the belief that only those with special personal qualities can dance the dance has one certain outcome: few employees will participate, and of those that do, nearly all will be highly paid professionals" (p. 94).

Job Characteristics

An alternative approach to selection is to focus primarily on the suitability of the job. This facilitates part-time telecommuting because tasks can be divided into those that are independent of worker location and those that are location-dependent (Kugelmass, 1995). Baruch and Nicholson (1997) identify two types of jobs that fit with homeworking: (a) technologically simple jobs with low autonomy which are easy to control remotely by focusing on outputs and (b) highly autonomous and professional positions where the work is complex and discretionary and control is via self-management. Other job characteristics to be considered are union membership, legal status, client/customer contact requirements, and security considerations regarding confidential or sensitive work materials (Estess, 1996).

To adequately assess job characteristics, management needs current and accurate information about job requirements. Job analysis is the process of gathering information about tasks, results, equipment, material, and work environment (Gatewood & Field, 1998). Since job analysis is performed in many organizations for compensation purposes, organizations could integrate telecommuting analysis into their standard job analysis process. This information also provides a basis for performance appraisals and for the design of effective telecommuter training.

Person-job fit is the critical factor in selection decisions but empirical data is still needed. Hartman et al. (1991) found no significant relationships between gender, education, age, occupational classification, nature of the job and telecommuter productivity or satisfaction.

Supervisor Characteristics

An increasing amount of attention is being given to profiling the effective telecommuting supervisor (Grensing-Pophal, 1999; Kugelmass, 1995; Leonhard, 1995; Mahfood, 1994; Nilles, 1998). Commerce Clearing House (CCH) guidelines (Telecommuting, 1999) divide supervisor characteristics into two categories: supervisor attributes and supervisory techniques. Attributes include: confidence of supervisory ability, belief that work is appropriate for telecommuting, desire to participate in telecommuting, proven ability to manage change, demonstrated ability to trust subordinates, and demonstrated flexible style that includes delegation, good communication, organization and computer literacy.

Trust and control issues are a major theme in discussions of supervisor attributes (Chapman et al., 1995; Grensing-Pophal, 1997; Nilles, 1998; Stavig, 1999). Even though studies indicate productivity increases when employees work at home (Dubrin & Bonard, 1993; Goldsborough, 1999), the supervisory fear is "if I can't see them, how do I know they are working?" (Kugelmass, 1995; Nilles, 1998). This psychological barrier is more important than technology and needs to be addressed in supervisory training programs (Davenport & Pearlson, 1998; Grensing-Pophal, 1997 & 1999).

One goal of the selection process must be a good fit between the employee and the supervisor. Telecommuter selection offers the opportunity to formally assess employee-supervisor fit and to evaluate it as a valid predictor of job performance. A poor supervisor can demotivate a good telecommuter. The best selection process can be subverted if supervisors are not trained, evaluated and rewarded for effective selection and support of telecommuters. Davenport and Pearlson (1998) recognize Hewlett Packard's efforts in taking a comprehensive approach to supervisory development.

Kugelmass (1995) questions the idea of a supervisor profile. He stresses that those who want to find a telecommuter persona also seemed compelled to find a specific supervisor persona and cites Connor's study finding no statistically significant relationship between management style and comfort with telecommuting. In that study both task- and person-oriented managers were comfortable supervising telecommuters but there is a definite need for more research on this topic.

Organizational Support Systems

One of the biggest problems with a mobile workforce is how to develop an organizational infrastructure that supports these efforts (Greengard, 1994). Two human resource management subsystems are integral supports for the telecommuter selection process: training and development, and performance appraisal (Gatewood & Field, 1998). In addition, the culture of the organization must support telecommuting (Baruch & Nicholson, 1997; Reilly, 1997).

Training programs. An organization's type and extent of training should be based on the level of employees' skills and abilities so that training programs complement the selection process (Gatewood & Field, 1998).

Piskurich (1996) divides effective training into five components: (a) a communication program that introduces telecommuting to the organization; (b) a component for interested employees to explore their suitability; (c) management training; (d) telecommuter skills training; and (e) training for non-telecommuting employees.

A training program can be part of the selection process. Merrill Lynch offers a two-week practice session for potential telecommuters. Employees work in a lab and communicate exclusively by telephone and e-mail. While this does not fully duplicate a home-based work situation, it does give a somewhat realistic job sample (Fusaro, 1997). This type of training builds employee skills and, in addition, it can help managers assess employees and also help employees self-assess their suitability for telecommuting.

Finally, training should include supervisors, as previously noted, and it is also appropriate and necessary for co-workers (Nilles, 1998; Vowles, 1996). Their understanding of telecommuting selection will enlist their support and assure them that it is not only the favored few who can telecommute, but that it is a business decision based on organizational goals and objectives.

Performance appraisal. Since the purpose of selection is to identify high performers, performance appraisal data can and should be used to evaluate the effectiveness of the selection process (Gatewood & Field, 1998). The organization's performance appraisal system should focus both the employee and the supervisor on outcomes, with telecommuters evaluated on results and not penalized for physical distance (Greengard, 1994). An organization that truly wants to encourage telecommuting must include performance objectives for managers to ensure their full support.

Organizational culture. A supportive organizational culture is necessary for successful homeworking in general (Baruch & Nicholson, 1997; Reilly, 1997) and it impacts the selection process specifically (Schuler & Jackson, 1996). Culture is the shared values, beliefs and assumptions in an organization that determine how individuals perceive, think about and react to their environment. Since top management sets the tone for corporate culture, its full support of telecommuting is critical (Nilles, 1998; Baruch & Nicholson, 1997). Comprehensive training programs help establish a supportive culture. An employee assistance program (EAP) can also help if EAP staff members are knowledgeable about the problems and opportunities of telecommuting. Newsletters and other employee communications need to explicitly demonstrate organizational support for telecommuters.

RELATED SELECTION ISSUES

A systematic approach to selection, used by well-trained managers, will address legal concerns in addition to selecting the best performers (Gatewood & Field, 1995). However, the selection of telecommuters often differs from an organization's standard process: Employees may request to telecommute in current positions; organizations may convert jobs and incumbent employees to telecommuting; people can gradually migrate into telecommuting; and telecommuting can be a retention strategy when a valued employee indicates an intention to leave. In these situations, an organization's selection process may be modified and it is easier to violate, or be perceived as violating, employment laws. Organizations should be aware of and educate supervisors on the potential legal pitfalls inherent in informal selection processes (Hartstein & Schulman, 1996). If telecommuting comes to be viewed as an employee entitlement (Nolan, 1999), more legal action on telecommuter selection decisions may be expected.

Because many telecommuting applicants are current employees it is easy to include them in the selection process. This helps employees identify barriers to telecommuting and encourages them to continue or to drop their pursuit of telecommuting. To facilitate self-assessment, employees can prepare a written proposal on: tasks that can best be done at home, time to be worked in telecommuting mode, communication plans, performance criteria, and other related issues (Estess, 1996). However, Baruch and Nicholson (1997) remind managers not to rely primarily on volunteers. It is to the organization's benefit to identify employees whose personal qualities and job characteristics are conducive to telecommuting.

Due to generally positive perceptions of telecommuting, organizations often need to inject reality into the selection process. The realistic job preview fully conveys both positive and negative aspects of a position and helps applicants determine whether they have inflated or overly optimistic expectations (Schuler & Jackson, 1996). Telecommuter self-help books can be given to prospective telecommuters early in the selection process and used as part of a realistic job preview (Bredin, 1996; Leonhard, 1995; Oberlin, 1997; Schepp & Schepp, 1995; Struck, 1995). The selection process can also include discussions with experienced telecommuters and managers who have supervised telecommuters. Mahfood (1994, p. 62) stresses that "It is incumbent on management to educate, encourage and inform the prospective telecommuter of the realities and potential problems of telecommuting."

Finally, human resource department personnel will be more effective in assisting managers if they are involved early in the selection process (Grensing-Pophal, 1998). They can help educate and counsel employees, managers, and co-workers, and can work to ensure legal compliance. Most importantly, they can take steps to ensure that telecommuting selection, and telecommuting programs in general, are supported by other human resource programs and activities (Greengard, 1994).

SUMMARY AND CONCLUSIONS

The conceptual framework presented here is based on a systems perspective and identifies critical variables and relationships in the telecommuter selection process. It is intended to inform and improve current telecommuter selection practices and to provide suggestions for future research.

A review of the telecommuting literature has found that several elements critical to successful selection are being addressed: employee characteristics, job characteristics, supervisor characteristics, and organizational support systems. What is lacking is a clear, empirical understanding of the elements, delineation of the relationships between the elements, and knowledge of how those relationships could improve telecommuter selection and, ultimately, telecommuter effectiveness.

Overall, the literature base on telecommuting still tends to be anecdotal, speculative, and lacks a strong research base as observed by Chapman et al. (1995). Switzer's (1997) annotated bibliography on telecommuting covers the available literature on management issues. McCloskey and Igbaria (1998) reviewed the empirical research on telecommuting and found many practitioner publications but only 32 published empirical studies with little or no investigation of the selection process. Future studies of telecommuting need to specifically address the telecommuter selection process for both internal and external candidates (Baruch, 1999). Given the projected increases in telecommuting, the systems framework for selecting telecommuters presented here provides guidance for organizational practice and future research.

REFERENCES

Apgar, M. IV. (1998). The alternative workplace: Changing where and how people work. *Harvard Business Review,* May/June, 76(3), 121-130.

Baruch, Y. (1999). *Teleworking: Benefits and pitfalls as perceived by professionals and managers.* Unpublished manuscript, University of East Anglia, Norwich.

Baruch, Y. & Nicholson, N. (1997). Home, sweet work: Requirement for effective home working. *Journal of General Management,* 23(2), 15-30.

Belanger, F. (1999). Workers' propensity to telecommute: An empirical study. *Information and Management,* 35(3), 139-153.

Bernardi, L. M. (1998). Telecommuting: Legal and management issues. *Canadian Manager,* Fall, 23(3), 18-20.

Bredin, A. (1996). *The virtual office survival handbook.* New York: John Wiley and Sons.

Briscoe, D. R. (1995). *International human resource management.* Englewood Cliffs, NJ: Prentice Hall.

Chapman, A. J., Sheehy, N. P., Heywood, S., Dooley, B., & Collins, S. C. (1995). The organizational implications of teleworking. In Cooper & Robertson (Eds.), *International Review of Industrial and Organizational Psychology,* 10(229-248). New York: John Wiley and Sons.

Davenport, T. H., & Pearlson, K. (1998). Two cheers for the virtual office. *Sloan Management Review,* 39(4), 51-65.

DiMartino, V., & Wirth, L. (1990). Telework: A new way of working and living. *International Labour Review,* September-October, 129(5), 529-554.

Dowling, P. J., Schuler, R .S. & Welch, D. E. (1994). *International dimensions of human resource management.* Belmont, CA: Wadsworth Publishing.

Dubrin, Andrew J., & Bonard, Janet C. (1993). What telecommuters like and dislike about their jobs. *Business Forum,* Summer, 18(3), 13-17.

Estess, P. S. (1996). *Working concepts for the future.* Menlo Park, CA: Crisp Publications.

Fusaro, B. (1997). How do we set up a telecommuting program that really works? *PC World,* 15(2), 238-240.

Gatewood, R. D., & Field, H. S. (1998). *Human resource selection.* Fort Worth, TX: Dryden Press.

Goldsborough, R. (1999). Making telecommuting work. *OfficeSystems99,* 16(8), 13.

Greengard, S. (1994). Making the virtual office a reality. *Personnel Journal,* September, 66-70.

Grensing-Pophal, L. (1999) Training supervisors to manage teleworkers, *HR Magazine,* January, 67-72.

Grensing-Pophal, L. (1997). Employing the best people from afar. *Workforce*, March, 30-38.

Grensing-Pophal, L. (1998). Training employees to Telecommute: A recipe for success. *HR Magazine*, December, 76-79.

Hamilton, C. (1987). Telecommuting. *Personnel Journal*, April, 91-101.

Hartman, R. I., Stoner, C. R., & Arora, R. (1991). An investigation of selected variables affecting telecommuting productivity and satisfaction. *Journal of Business and Psychology*, 6(2), 207-225.

Hartstein, B. A. & Schulman, M. L. (1996). Telecommuting: The new workplace of the 90s. *Employee Relations Law Journal*, Spring, 179-88.

Kugelmass, J. (1995). *Telecommuting*. New York: Lexington Books.

Legal and Contractual Situation of Teleworkers. (1997). *The Social Implications of Teleworking*. Dublin, Ireland: European Foundation for the Improvement of Living and Working Conditions.

Leonhard, W. (1995). *The underground guide to telecommuting*. Reading, MA: Addison-Wesley.

Mahfood, P. (1994). *Managing the home based worker*. Chicago: Probus Publishing.

McCloskey, D.W. & Igbaria, M. (1998). A review of the empirical research on telecommuting and directions for future research. In M. Igbaria & M. Tan (Eds.), *The Virtual Workplace*. Hershey, PA: Idea Group Publishing.

McCune, J. C. (1998). Telecommuting revisited. *Management Review*, 87(2), 10-16.

McGonegle, K. (1996). Taking work off-site: AEGON's telecommuting program. *Employment Relations Today*, Spring, 23(1), 25-38.

Moorcroft, S. & Bennett, V. (1995). *European guide to teleworking: A framework for action*. Dublin, Ireland: European Foundation for the Improvement of Living and Working Conditions.

Messmer, M. (1998). *The fast forward MBA in hiring*. New York: John Wiley and Sons.

Nilles, J. M. (1994). *Making telecommuting happen: A guide for telemanagers and telecommuters*. New York: Van Nostrand Reinhold.

Nilles, J. M. (1998). *Managing telework: Strategies for managing the virtual workforce*. New York: Wiley.

Nolan, P. (1999). Happily ever after. *Twin Cities Business Monthly*. September, 95-100.

Oberlin, L. H. (1997). *Working at home*. Franklin Lakes, NJ: Career Press.

Piskurich, G. (1996). Making telecommuting work. *Training and Develop-*

Pynes, J. E. (1997). *Human resources management for public and nonprofit organizations*. San Francisco: Jossey-Bass.

Reilly, E. M. (1997). Telecommuting: putting policy into practice. *HR Focus*, 74(9), 5-7.

Schepp, B. & Schepp, D. (1995). *The telecommuter's handbook: How to earn a living without going to the office*. New York: McGraw-Hill.

Shuler, R. S., & Jackson, S. E. (1996). *Human resource management*. Minneapolis/St. Paul, MN: West Publishing.

Stavig, V. (1999). The star treatment. *Twin Cities Business Monthly*. September, 89-91.

Sudbury, D. A., & Towns, D. M. (1997). Traps for the unwary: Employment law implications of telecommuting. *Employee Relations Law Journal*, 23(3), 5-30.

Struck, N. (1995). *Working smarter from home*. Menlo Park, CA: Crisp Publications.

Switzer, T. R. (1997). *Telecommuters, the workforce of the twenty-first century: An annotated bibliography*. Lanham, MD: Scarecrow Press.

Telecommuting: Selecting staff. (1999). Commerce Clearing House-EXP, HRM-Personnel@2437.

Vowles, A. (1996). Home is where the office is. *CMA Magazine*, 70, 19-23.

Chapter VIII

Managing the Virtual Team: Critical Skills and Knowledge for Successful Performance

Richard G. Platt and Diana Page
University of West Florida, USA

ABSTRACT

Expectations for manager and employee workplace relationships are changing because of telecommunications technology. Telecommuting allows organizations to effectively combine and use the skills and knowledge of off- and on-site employees. The focus of this chapter is to describe the unique environment and problems presented by virtual teams and to outline the skills and knowledge employees and managers must have so they can achieve team and organizational goals.

Two or more people working together to achieve the same objective, but geographically separated and electronically connected, is a simple definition of a virtual team. (Lipnack & Stamps, 1997). Virtual teams can share understandings, facilitate the exchange of ideas, enhance communication and encourage continuous learning. Virtual teams allow organizations to build effective teams from personnel who might not otherwise be available to work together, enhance the availability of resources from outside the organization, hire and retain the best people regardless of location, and gain access to needed expertise (Townsend, DeMarie, & Hendrickson, 1996).

There are several trends that will escalate the growth of virtual teams in 21st-century organizations: the prevalence of flat organizational structures, environments that require cooperation and competition among organizations, changes in workers' expectations of organizational participation, the shift toward service/knowledge environments, and increasing globalization of trade activities (Townsend, DeMarie, & Hendrickson, 1998, p.18). These trends create new challenges for virtual teams compared to the challenges facing traditional teams. Knowledge of the organization, technological capabilities, and mastery of effective management techniques in this emerging environment are core skills for virtual team success. Telecommuting teams must adjust to this new environment by learning more about 1) how the organization functions, 2) how to use new technology productively, and 3) how to manage and promote effective management habits.

KNOWLEDGE OF THE ORGANIZATION AND ITS GOALS

Restructuring, or changing an organization's structure in an attempt to improve and enhance performance, is an ongoing management activity. Changing external and internal environments and technological innovations offer a motivation for continuous organizational rearrangements. The functional and divisional structures that dominated management history have morphed into more hybrid structures that emphasize teams, networks, fewer levels of management, improved communications technologies, wider spans of control, and effective delegation and empowerment techniques. Both managers and employees must recognize how these factors impact virtual team performance.

Successful virtual teams will take their direction from the way a business is organized and how it achieves its goals. Organizational mission, goals, and communication channels are the elements which provide critical links between teams and their organization. The structure of an organization describes jobs and relationships, who is supposed to do what, who is responsible for what, and how different people and organizational units are related to one another. Thus, understanding how organizations are structured is important for any team. Function, product, customer, geographic and matrix structures were created to establish appropriate communications and coordinate efforts to achieve organizational goals. These traditional structures worked well until the explosion of new technology, globally oriented businesses, and heightened competition impacted organizations. Today's organizational structures are evolving toward team based, autonomous organizational structures (Dess, Rasheed, McLaughlin & Priem, 1995; Dwyer, 1994; Halal, 1994; Morhman, Cohen, & Mohrman Jr., 1995) as a result of these influences.

Understanding these emerging organizational structures is a key element that will better prepare employees and managers to establish and design effective virtual teams. Virtual teams facilitate coordination and add flexibility to organizations. Thus, they provide one of the major links that allow organizations to transform themselves from traditional hierarchical structures into 21st-century architectures. Descriptions of three of the more popular evolving organizational architectures and their implications follow.

Modular Organizations

Modular organizations include a limited, relatively small core of permanent employees and facilities. Because of the small number of employees, interactions occur on an as-needed basis rather than through the formal channels and structures used in larger organizations. Managers in modular structures outsource a number of organizational functions, and hire outside contractors, production and service employees to support the business. These employees generally are connected through computers and electronic communication devices. In the modular organization individual employees may be part of more than one team.

For employees who desire autonomy and independence, modular organizations may be ideal, but employees who want more structure may feel isolated and unconnected to the organization. Managers and team members may be uncomfortable relying on others about whom they know little but whose participation and reliability are necessary to complete needed tasks and activities. Additionally, virtual team employees may be part-time or contract employees. As such, these employees will be paid for performance, but typically will not receive organizational, medical and retirement benefits. This practice favors the organization, but may not be seen as a benefit to the virtual employee.

Virtual Organizations

A virtual organization is a continually evolving alliance of independent companies, including suppliers, customers, and competitors, that is linked together to share skills, costs and access to each others' markets. (Dess, Rasheed, McLaughlin, & Priem 1995). These structures need not be permanent and may not have a central office. Participating organizations may be involved in several alliances at any given time. These alliances may cause organizations to function as if they have more capacity and resources than they actually possess. What these organizations lose in strategic control, they gain in a collective strategy that allows them to cope with uncertainty and adapt to current trends through cooperative efforts. Virtual team members employed with virtual organizations can enhance their performance by understanding the missions and goals of each of the alliances within the virtual organization and understanding how the organizations are associated with one another. Moving among the virtual organizations' alliances, virtual team members must demonstrate flexibility, excellent communication skills, and knowledge of the industries involved.

Boundaryless Organizations

Boundaryless organizations are created to improve task coordination and flexibility, thus making the organization more responsive to environmental changes. The boundaryless structure replaces departments with cross-functional teams to improve horizontal relations. Cross-functional teams are composed of members from different areas of work responsibility such as marketing, finance, and operations. Even if these teams operate solely within a single organization, the size and location of the organization may require them to act in a virtual environment. As the use of organizational teams increases, control shifts from managers to employees. Again, more responsibility will be thrust upon virtual employees, particularly for effective communications. If the necessary authority accompanies those responsibilities, employees and organizations will benefit.

Bureaucracies are not likely to disappear quickly, but using virtual teams can help organizations transition partially or wholly to team-based structures. Virtual teams can provide the crucial links that connect different units within an organization and thereby helping to change the structural shape of the organization and bring about improved organizational performances (Applebaum & Blatt, 1994; Kalleberg & Moody, 1994; Levine & D'Andrea, 1990). Virtual teams that form the hub of future organizational structures must recognize and create relational links that strengthen communications throughout the organization.

Evolving Reporting Relationships

Computer-mediated communication that allows employees easy access to managers at all organizational levels is blurring the traditional principles of organizing. The chain of command, the line of authority that vertically links all employees with successively higher levels of management, is contracting. Unity of command, the management principle that each person in an organization should report to one and only one supervisor, mutates as the number and variety of virtual teams increase. As virtual teams evolve and learn to manage themselves, hierarchical interdependencies dissolve, allowing teams to develop stronger interdependencies with other teams and organizations. Span of control, the number of persons reporting directly to managers, widens as organizations evolve toward the new architectures. Thus, managers must use appropriate delegation and empowerment techniques to transfer authority and responsibility to employees and virtual teams.

The steps in this delegation process are fairly straightforward: 1) define the responsibility, 2) agree on performance objectives and time frames, 3) develop trust, 4) provide task support, 5) give performance feedback, 6) help when things go wrong, and 6) retain accountability. In practice, however, delegation is more difficult than it appears. Developing trust (Handy, 1995) and giving performance feedback (Armstrong & Cole, 1995) are blocked when distance and time separate managers and employees.

Supervising and monitoring performance in the past could be accomplished by walking around and personally checking on activities. For managers of some telecommuting teams, "management by walking around" means traveling to different sites to visit team members. For other managers, "management by walking around" means involvement in team member conversations and maintaining communications connections for multiple remote users. Paradoxically, the more virtual the relationships, the stronger the need for face-to-face contact (Handy, 1995).

Required organizational knowledge for virtual team members includes an understanding of the organization's strategic plan, mission and goals, and the principles that define formal organizational relationships. Lipnack and Stamps (1999) suggest virtual employees and teams must clearly identify why they exist and translate their purpose into action steps that become the basis for the work they do together. Doing this will aid the team in defining their value added. Cooperation, interdependence and concrete results are expected outcomes for virtual teams. Further, managers will expect telecommuters to solve many of the technological glitches that might prevent regular communications and impair the production process.

TECHNOLOGY AND TECHNOLOGICAL COMPETENCIES FOR VIRTUAL TEAMS

Mastering technological complexities is the next important element required to work effectively in virtual teams. Notebook computers, high-speed modems, fax machines, virtual private networks, and groupware are just a few of the technologies used to gain the expertise of off-site workers who cannot, or will not, travel to central offices to work. Harnessing these technologies to build cohesive, productive teams is an increasingly difficult task and a critical managerial challenge.

Since the introduction of the IBM PC in 1981, increasingly powerful desktop computers and analytical software have come within the reach of individual workers. With the advent of desktop and portable microcomputers, workers no longer had to wait for technicians in white coats to produce reams of printouts from data processing centers. Instead, following the predictions of Moore's Law, microcomputers continued their unwavering evolution of becoming faster, cheaper and more powerful until the computing power of yesterday's mainframe is today economically available on the desktop or briefcase.

Software functionality likewise increased. From the rudimentary capabilities of the earliest file management systems and spreadsheets, software evolved into user friendly tools for solving complex decision support problems. Today, office application suites integrate data analysis, document development, and presentation tools for storing, processing and analyzing complex business tasks for less than the price of a tailored suit or a short vacation.

With the ready availability of all this inexpensive technology, initially the assumption followed that workers enabled by all this technology would be more productive than workers without technological assistance. What followed was a rush to invest in technology, motivated by the anticipated benefit of corresponding increases in white-collar productivity. However, the resulting increase in productivity never materialized as originally predicted. Instead, despite the investment of billions of dollars in technology in the 1980s, there were no corresponding measurable increases in white-collar productivity. Nobel-winning economist Robert Solow (1987, p. 36) branded this inconsistency as the "productivity paradox," a discrepancy between the expected economic benefits of computerization and the actual measured effects.

In the 1990s this inconsistency began to change. The convergence of hardware, networking, and collaboration technologies totally changed the landscape of computing and collaboration. These changes empowered telecommuters and virtual teams, improved white-collar productivity, and reversed the productivity paradox. Verification of this reversal was offered by Federal Reserve Chairman Alan Greenspan (1999) when he reported non-farm business productivity gains over the past years, approaching two and one-half percent, caused in large part to advances in the synergy resulting from the restructuring of business processes combined with the innovations in new technologies.

Just as in business restructuring and reengineering, technology is an enabling factor—not the driving factor—in creating the virtual team. Advances in hardware, software and communications provide the infrastructure to engender virtual teams. This infrastructure can link anyone to everyone, allow the sharing of information, and, even more importantly, facilitate collaboration. The technology allows the telecommuting team member to stay in contact with the work group while out of the office, and to collaborate with the work group on team activities and work products. The following sections describe the supporting technologies that assist virtual team members and managers.

Hardware Technology

Many of today's microcomputer systems look very similar to the 1981-vintage IBM PC on the outside, but on the inside there are astonishing differences. The microprocessor chip packs more than five million semiconductor circuits and an internal speed more than 100 times faster than the original IBM PC. Operating systems provide user-friendly graphical interfaces and user-friendly help systems that diagnose simple malfunctions and automatically install new peripheral components. Secondary storage includes hard disk drives that store billions of bytes and optical drives that can store an entire full-length movie, or even a dynamic, interactive sales presentation. Zip® disk removable storage media allow 100 megabytes and more to be carried in a flat package that fits into a shirt pocket.

Take all these developments, then shrink them into a package the size of a single volume of an encyclopedia. Replace the bulky monitor with a flat, LCD panel display. Add to the system a modem card, which can communicate at speeds of 56,000 bits per second, reduce the weight to under seven pounds and the notebook computer, tool of the mobile road warrior and telecommuter, is created.

Today's computers are easier to set up and maintain. The telecommuter's office once was a maze of cables connecting computers, printers, scanners, and fax machines that only a well-trained technician or the adventuresome novice attempted to assemble. Today, vendors simplify equipment setup by color coding both ends of connecting cables and provide free technical support for those instances when even color coding is insufficient. Even maintenance is made easier by the emergence of user-replaceable, disposable components.

With this new maintenance philosophy comes an added responsibility, which is especially important for workers who are not colocated with the technical support staff. In 1995 telecommuting workers might have expected their organization to provide the initial hardware configuration and setup at a remote location. Today, increased expectations are made of remote workers. They must know how to connect their computers to telephone systems using a variety of ingenious methods. For example, how does a traveling worker connect a laptop modem to the telephone system in a hotel that does not provide data ports and where the telephone does not have a removable jack connection? Resolving this situation does not require a permanent technical assistant, but it does require that today's remote workers have more knowledge of the technology—both hardware and software—than did their predecessors.

Virtual Team Communications Challenges

For years, organizations built sophisticated communications networks to support organizational systems based on the exclusivity factor. Network designers did not consider the public communication networks in their design alternatives. Network designs were based on the internal organizational communications requirements that translated into proprietary networks running over leased lines (Gasparro, 1997). However, these communications systems were predicated on communications between fixed locations with no expectation of remote workers. Neither telecommuting workers nor virtual teams were on the planning horizon.

When most people think of connectivity for remote workers or virtual work teams, the first thought is often of laptops, modems, and phone lines. However, the ability of a computer to receive information is constrained because current telephone company technology puts digital data into analog form for transmission over existing telephone lines, only to require modems to change it back into digital form. In other words, remote worker collaboration is hampered by the bottleneck caused by the limited bandwidth of analog data transmission over the existing, voice-grade telephone infrastructure. To fully enable telecommuting workers and virtual teams, alternatives to reduce or eliminate the analog bottleneck must be identified and explored.

One alternative to the traditional copper wire technology is ISDN, the integrated services digital network. ISDN is an international communications standard for sending voice, video, and data over digital telephone lines at transfer rates of 128 Kb/s. This is more than twice the data rate provided by today's fastest analog modems. Accessing ISDN services requires special modems, telephones, fax machines, and other equipment that are ISDN compatible, but these are usable only with an ISDN telephone line— something that is expensive and not universally available.

Another alternative to analog data transmission is asymmetric digital subscriber line technology (ADSL). ADSL increases the data transfer rate between the home or office and the telephone company switching stations. Asymmetric transmission means upstream (user to telephone company) and downstream (telephone company to user) data transfer rates are not the same. ADSL supports data rates of from 1.5 to 9 Mb/s when receiving data (known as the downstream rate) while upstream rates vary from 16 to 640 Kb/s when sending data. Having a faster downstream data transfer rate is highly beneficial because the highest volumes of data transmissions are the downstream transmissions received by the user. Another benefit is that ADSL furnishes three channels, two for data and one for voice. Thus the telecommuter can work on the computer and simultaneously talk on the telephone without having to lease two telephone lines.

Both ISDN and ADSL require special modems and service from a telephone company. Although ADSL is a relatively new technology, it is available in most major metropolitan areas and is spreading rapidly. Therefore, ADSL can be a major advantage when the virtual team member works from a fixed location (e.g., remote office), but is of minimal benefit to the telecommuting worker. As demands for faster Internet access increase, telephone companies are moving to extend ADSL availability. Soon, road warriors will have ADSL connectivity available in hotels, airports, and other travel stops.

Regardless of the type of organization, maintaining communications connections for multiple remote users is both costly and time intensive. Initially, supporting one remote user required the organization's infrastructure to provide a telephone line with a modem and telecommunication software. Maintenance of this one dial-up connection was rather simple, but as the number of remote users increased, the cost and complexity of dial-up connections increased rapidly.

As more workers needed remote access to organizational systems, the first solution was an extensive dial-up infrastructure. However, having more users does not require maintaining a one-to-one relationship between the number of remote users and the number of telephone lines because not everyone will need to be on-line at the same time. This presents a serious dilemma for management. The telecommunications manager must constantly balance effective service in the face of increasing demand while simultaneously controlling infrastructure expenditures; too many telephone lines waste money and too few inhibit remote worker capability and effectiveness. Building the infrastructure requires building banks of modems, installing the required incoming type (e.g., ADSL, ISDN, analog) and number of telephone lines, and the purchase of telecommunications software. Maintaining that infrastructure is an ongoing and increasingly expensive process. The telecommunications infrastructure requires the constant attention of expensive telecommunications technicians, replacement of parts, and the ongoing cost of leasing lines.

The virtual private network (VPN) is a communications tool that allows remote workers to access organizational information systems as if they were connected directly to the organization's internal computer network, but a VPN eliminates the need for a dial-up infrastructure. Further, a VPN provides the organization a secure, cost effective technique to leverage its organizational systems to persons from other external organizations.

A VPN is exactly what the name implies; a private data network that really is not there. The VPN uses the existing, public Internet infrastructure by building a virtual tunnel for passing secure communications. The VPN tunnel maintains the privacy of communication by using a tunneling protocol and encryption. To use a VPN, a message is encrypted before it is sent through the public data network and then decrypted at the receiving end, making the public network act like a private network. The VPN gives an organization the same networking capabilities as a dial-up infrastructure at a much lower cost. Remote users connect via any commercial ISP and use the public Internet infrastructure rather than requiring the organization to maintain a private one. In short, the VPN gives remote users immediate access from any point in the world, through any ISP, to any organizational network application, with organization-controlled security and authentication and minimal communication costs. The burden of maintaining the dial-up infrastructure is transferred from the organization to the ISP.

Groupware-Software for Collaboration

Telecommuting only requires one person working at a distance. For more than 15 years telecommunications technologies have allowed individuals to work independently at a distance. However, these same workers have not had collaboration tools that enable two distant workers to work together on the same project.

In contrast, virtual teams by definition require at least two team members working at a distance. These team members must not only be able to communicate, they must also be able to collaborate. Today's groupware applications allow team members at different locations to simultaneously share an application and concurrently collaborate to share ideas and efforts on the same project without having the delay of transmission from one location to another, not even the delay associated with e-mail. In a virtual team, members not only have to communicate, they must have the tools and technologies to collaborate.

Groupware refers to computer applications that support group activities by allowing group members to work together by sharing applications, databases, common bodies of text, colleagues' schedules, etc. Groupware has been marketed with an emphasis upon shared effort, collaboration, and cooperation. Lotus Notes, for example, lets groups located in different offices and locations share messages and memos via computerized bulletin boards and document databases.

Groupware spreads power far more widely than before by giving lower-level workers access to information previously unavailable or restricted to upper management. Many who have grown comfortable near the top of the information chain are understandably distressed by the way groupware disrupts old-style hierarchies. The new groupware tools are so powerful they virtually compel companies to reengineer themselves into knowledge organizations (Kirkpatrick, 1993).

Groupware is not restricted to e-mail or video conferencing, although these are two tools that the virtual team may use. Instead groupware consists of application such as Lotus Notes, Microsoft NetMeeting, AOL Instant Messenger, ICQ, and TeamWave. Each of these tools provides different forms of collaboration from screen sharing to application sharing to inexpensive audio and video conferencing, all via the Internet.

Entire web sites are devoted to providing services designed to support team collaboration, many of them free. For example, http://www.intranets.com provides a simple, group collaboration tool (note board, group e-mail, calendar, etc.) at no charge. More of these tools appear every day. The question is not whether employees will work on a virtual team, but which collaboration tools will employees choose?

MANAGING VIRTUAL TEAMS

Working in teams, once considered a futuristic best way to get things done in organizations, now is a necessary skill expected of most managers and employees. The popularity of teams and team culture in organizations originated in the 1980s. Prior to the 1980s, the literature on teams was sparse. Earlier management assumptions that employees placed on organizational teams willingly would change their behaviors to accomplish goals, proved untrue. Thus, as the use of teams exploded in the 1990s, managers were forced to seek ways to improve team performance. Researchers investigated many team variables and identified specific behavioral elements present in successful team activities. Bettenhausen's (1991) report of more than 250 studies about group research between January 1986 and October 1989 places group research findings into three categories: group dynamics, group interaction (e.g., with structure, technology, environment), and social psychology influences (e.g., polarization, social loafing, goal setting).

Characteristics of High Performance Teams

High performance teams must recognize the stages of group development, understand team roles, and build trust. Virtual team members must not only be familiar with group dynamics concepts, but must also be able to use these concepts within the technological environment where virtual teams reside. Thus, it is important for team members to understand a few basics about how successful teams operate. The key to developing effective group norms and culture begins by understanding the stages of group development: forming, storming, norming, performing, and adjourning. These five stages (Tuckman, 1965) describe how group members relate interpersonally. All groups go through a developmental process, but the amount of time each group spends in each stage varies. Examples of variables that can effect the stage of group development include new members or a change in goals. These influences may cause the development to regress or repeat.

Team Roles

Understanding team roles and eliminating role ambiguity are important elements in managing virtual teams, and they deserve priority as the team is formed. Successful teams need a balance of task and relationship roles;

therefore, team member selection should consider this balance (Bales, 1958; Benne & Sheats, 1948). Task roles help the team complete specific activities and achieve its goal, whereas relationship roles refer to the social interaction that helps team members bond and work in harmony. Achieving goals comes with balancing these roles, and that balance differs with team members, the task, the environment, and the project deadline.

Developing relational roles is more difficult and takes longer to achieve for virtual teams. Relational roles are significant contributors to effective information exchange (Warkentin, Sayeed, & Hightower, 1997). Because exchanging social-emotional information exemplified in relational roles is difficult, virtual teams tend to be more task oriented and neglect important relationship-oriented information (Chidambaram, 1996). Some positive outcomes associated with relational links include enhanced creativity and motivation, increased morale and improved decisions (Walther & Burgoon, 1992). The reduced relationship building that occurs in virtual teams implies the use of traditional meetings as a supplement to electronically connected teams would be helpful (Warkentin, et al, 1997).

Achieving organizational goals more likely will be accomplished when team members cooperate with each other and share their resources and expertise. Where mergers, streamlining, and change are everyday activities, it is common for employees to compete with each other and to fear a loss of status. At a time when sharing information and resources is critical, cooperation among team members may be most difficult to achieve. Without such cooperation, teams will fail to achieve their individual as well as organizational goals.

Since trust building and negotiating activities occur in the early stages of group development, an initial face-to-face period when team members get to know each other is important (Jarvenpaa, Knoll, & Leidner, 1998). This understanding accelerates group productivity and dramatically reduces the communication effort required at a distance.

Building Trust

Building trust includes open communication, integrity, and competence. Open communication requires team members to act in the present moment. Technology that employs synchronous electronic media can enhance virtual teams capability for openness. However, virtual teams whose interactions are communicated electronically miss the valuable face-to face cues (e.g., facial expressions, gestures, and vocal inflections) easily observed in traditional team meetings (Daft & Lengel, 1984; Daft, Lengel & Trevino, 1987). Without these cues communication is more difficult, and participants must reconstruct communication patterns in ways that give them the information these cues would provide. For example, distant employees tend to be forgotten, remote employees are unable to observe valuable role models (Armstrong & Cole, 1995) and have difficulty gaining group member consensus (McLeod, 1992). Solving conflicts and gaining commitment are more easily accomplished in face-to-face meetings (DeMeyer, 1991; Hollinghead & McGrath, 1995). Building trust in virtual teams is expressed in several ways.

Significant variables which describe virtual teams that have developed trust are: Team members volunteer for roles, exhibit individual initiative, deal decisively with free-riders, focus on results, shower other members with encouragement and support, rotate leadership, spend considerable time discussing goals and time constraints, give substantial feedback toward improving content of colleague's work, and exhibit empathetic task behavior (Jarvenpaa, Knoll & Leidner, 1998). Temporary teams, members who have never worked together and are unlikely to work together again, may operate a little differently. For temporary teams, Myerson, Weick & Kramer, (1996) suggest "swift trust" can emerge which involves more depersonalized actions. "Swift trust," whereby team members act as if trust is present, enables members to take action. In turn, this action aids team members in dissolving uncertainty, ambiguity and vulnerability.

One final but important tool for virtual groups is a psychological contract: a set of expectations about who will provide what in the relationship. The contract is dynamic in nature, thus it may be updated or changed as desired. As the name implies, psychological contracts are generally implied understandings however, we suggest teams develop written psychological contracts. Thus, groups can spell out group roles, attain goal clarity, establish positive group norms and group behavioral expectations.

SUMMARY

Successful virtual teams do not happen just because someone gets a computer, a modem and a yen to work with someone at a distance. Successful virtual team management requires that the team manager understand the significance and interaction of the ingredients that make virtual teams productive—the Six Ts of Virtual Teams: team formation, training, task definition, telecommunication, tools, and trust.

Just like face-to-face teams, virtual teams evolve through the developmental stages of forming, storming, norming, performing, and adjourning.

The successful virtual team manager purposefully steers team members through these stages of team formation. It is important to assemble the virtual team in a face-to-face meeting during this team-building stage of group development. Meetings of this type are proven methods of establishing both relational links (Warkentin, 1997; Walther & Burgoon, 1992) and trust (Jarvenpaa, 1998), which are critical to virtual team success.

Training implies instructing the members of virtual teams not only on how to use the new technologies, but also how to adapt to new organizational structures and goals. These new organizational structures are shifting away from hierarchies with strict reporting relationships toward hybrid structures in which team membership spans organizational boundaries and blurs static reporting relationships. These new patterns also transform employee-employer relationships. A major difference is the way in which performance is measured in virtual teams. The virtual team manager must be prepared to evaluate team members on task performance without direct observation of work habits and patterns.

Successful virtual team performance requires the elimination of ambiguity through the explicit definition of task objectives and clear establishment of team member roles and responsibilities for those objectives. The team leader will need to clarify organizational standing plans and how these correlate with task objectives. Familiarity with these planning concepts places virtual teams in a better position to produce the results desired by contracting organizations. A clear definition of the tasks and roles helps the manager understand the patterns of individual behavior. Task-oriented members concentrate on getting the job done; relationship-oriented members concentrate on the social interaction of team members. Managers who drive employees' concentration to the "bottom line" or to production-only issues will find important quality issues lacking in project results. High-performance teams must balance task and process behaviors to increase the likelihood of quality project results.

Telecommunications is the underpinning of virtual teams; the technology that builds the framework and allows one worker to communicate with another at a distance. It allows extensive person-to-person links, information sharing and real-time collaboration. Virtual team technology includes both the computing devices and the communications networks necessary to provide the infrastructure for virtual team to operate. The virtual team manager must understand the hardware requirements for individual team members and the network alternatives necessary to build the virtual environment that unites the team and within which it will operate. Together, these technologies equip the team members with the foundation for team success.

The tools for virtual team success are the groupware collaboration tools that enable the team to work together as a unit. These groupware tools let multiple group members work together on the same project component by supporting shared effort, collaboration, and cooperation. These tools allow a team member at one location to concurrently work with other team members at other locations; sharing ideas, cross-pollinating creativity, distributing effort, and all the other collaborative activities that contribute to task accomplishment.

Virtual teams must develop trust among their members since they operate in an environment dependent on the performance of persons outside their sphere of influence. To gain "swift trust" (Meyerson et al. 1996), virtual teams focus more attention on cognitive issues and less on social interaction. Developing psychological contracts will help teams establish positive group norms and group behavioral expectations that foster trust.

This chapter discussed the core skill and knowledge (the six Ts) needed for managing the virtual team to successful performance. These core skills allow virtual teams to fill an important labor gap for organizations and they create an opportunity for entrepreneurial employees to define their own work parameters.

REFERENCES

Applebaum, E., & Blatt, R. (1994). *The new American workplace*. Ithaca, NY: ILR.

Armstrong, D. J. & Cole, P. 1995. Managing distances and differences in geographically distributed work groups. In S. E. Jackson & M. N. Ruderman (Eds.), *Diversity in workteams: Research paradigms for a changing workplace*. Washington, D.C.: American Psychological Association.

Bales, T. (1958). Task roles and social roles in problem-solving groups. In E. E. Macoby, T. M. Newcomb, & E. I. Hartley (Eds.), *Readings in Social Psychology*. New York: Holt, Rinehart, & Winston.

Bettenhausen, K. L. (1991). Five years of group research: What we have learned and what needs to be addressed. *Journal of Management,* 17(2), 345-381.

Benne, K. D., & Sheats, P. (1948). Functional roles of group members. *Journal of Social Issues*, 41-49.

Chidambaram, L. (1996). Relational development in computer-supported groups. *MIS Quarterly,* 20(2), 143-163.

Daft, R. L., & Lengel, R. H. (1984). Information richness: A new approach to managerial behavior and organizational design. *Research in Organizational Behavior,* 6, 191-223.

Daft, R. L., Lengel, R. H., & Trevino, L. K. (1987). Message equivocality, media selection, manager performance: Implications for information systems. *MIS Quarterly,* 11(3), 355-366.

DeMeyer, A. (1991). Tech talk: How managers are stimulating global R & D communication. *Sloan Management Review,* 32(3), 49-58.

Dess, G., Rasheed, A. M. A., McLaughlin, K. J., & Priem, R. (1995). The new corporate architecture. *Academy of Management Executive,* 9(3), 7-20.

Dwyer, P. (1994). Tearing up today's organizational chart. *Business Week,* (11), 81-82.

Gasparro, D. (May, 1997). Charting the data VPN movement. [On-line] http://www.teledotcom.com/0597/pl/tdc0597plvpn.corp.html.

Greenspan, A. (September 8, 1999). Maintaining economic vitality. Millennium Lecture Series, Grand Rapids, Michigan. [On-line] http://www.bog.frb.fed.us/boarddocs/speeches/1999/19990908.htm

Halal, W. E. (1994). From hierarchy to enterprise: Internal markets are the new foundations of management. *The Executive,* (11), 69-83.

Handy, C. (May-June 1995). Trust and the virtual organization. *Harvard Business Review,* 73(3), 40-50.

Hollingshead, A. B., & McGrath, J. E. (1995). Computer-assisted groups: A critical review of the empirical research. In R. A. Guzzo & E. Salas (Eds.), *Team Effectiveness and Decision Making in Organizations.* San Francisco: Jossey-Bass.

Jarvenpaa, S. L., Knoll, K., & Leidner, D. E. (1998). Is anybody out there?: Antecedents of trust in global virtual teams. *Journal of Management Information Systems,* 14(4), 29-64.

Kalleberg, A. L. & Moody, J. W. (1994). Human resource management and organizational performance. *American Behavioral Science,* 37(9), 48-62.

Kirkpatrick, D. (1993). Groupware goes boom. *Fortune,* 128(16), 99-103.

Levine, D. I. & D'Andrea, T. L. (1990). Participation, productivity, and the firm's environment. In A. S. Blinder (Eds.), *Paying for productivity.* Washington, DC: Brookings Institute.

Lipnack, J., & Stamps, J. (1997). *Virtual Teams, Reaching Across Space, Time, and Organizations with Technology.* New York: John Wiley &

Sons, Inc. p. 6.

Lipnack, J. & Stamps, J. (1999). Virtual teams: The new way to work. *Strategy & Leadership,* 27(6), 14-20.

McLeod, P. L. (1992). An assessment of the experimental literature on electronic support of group work: Results of a meta-analysis. *Human Computer Interaction,* 7(3), 257-280.

Meyerson, D., Weick, K. E., & Kramer, R. M. (1996). Swift trust in temporary groups. In R. M. Kramer & T. R. Tyler (Eds.), *Trust in organizations: Frontiers of theory and research.* Thousand Oaks, CA: Sage Publications.

Mohrman, S. A., Cohen, S. G., & Mohrman Jr., A. M. (1995). *Designing team-based organizations.* San Francisco: Jossey-Bass.

Solow, R. M. (1987). We'd better watch out. *New York Times Book Review,* July 12, p. 36.

Townsend, A. M., DeMarie, S. M., & Hendrickson, A. R. (1996). Are you ready for virtual teams? *HRM Magazine,* 9, 123-126.

Townsend, A. M., DeMarie, S. M. & Hendrickson, A. R. (1998). Virtual teams: Technology and the workplace of the future. *Academy of Management Executive,* 12(3), 17-29.

Tuckman, B. W. (1965). Developmental sequence in small groups. *Psychological Bulletin,* 63(4), 384-399.

Walther, J. B., & Burgoon, J. K. (1992). Relational communication in computer-mediated interaction. *Human Communication Research,* 19(1), 50-88.

Warkentin, M. E., Sayeed, L., & Hightower, R. (1997). Virtual teams versus face-to-face teams: An exploratory study of a web-based conference system. *Decision Sciences,* 28(1), 975-996.

Chapter IX

Case Study of the St. Paul Companies' Virtual Office for the Risk Control Division

Nancy J. Johnson
Capella University, USA

EXECUTIVE SUMMARY

The St. Paul Companies has successfully implemented a virtual office (VO) working environment for its construction risk control and commercial risk control employees over the past six years. The program goals of operating more cost-effectively, increasing contact of the risk control specialists with their customers, and reducing office space costs for The St. Paul Companies have been met. There are many good practices that have been developed over the six years of offering the program, and more refinements and changes planned. As the communications and computer technologies advance, facilitation of working from remote sites improves. While it is easier for employees to work from remote sites, maintaining the boundaries between work and personal lives is more challenging. Improving the VO employees' and corporate employees' understanding of the others' working conditions is necessary to improve relationships and the acceptance of change. The concept of VO work is well established within the organization, and the demand for it is growing.

BACKGROUND

Founded in 1853, The St. Paul Companies is Minnesota's oldest business corporation. It ranks 171 on the Fortune 500 list of the largest companies in the US. The multi-line, worldwide insurance company consists of several separate entities, and is growing continuously through mergers and acquisitions. The St. Paul Fire and Marine Insurance Company is the US-based property-liability insurance underwriting operation, with the St. Paul International Underwriting encompassing the rest of the world operations. F & G Life is the US life insurance underwriting and annuity operation, and the asset management business is accomplished through a majority ownership of the John Nuveen Company (St. Paul Companies, 2000). For 1999, the operating earnings of the firm increased to $636.3 million. The CEO credited the improvements to the intense restructuring efforts to transform the general property-casualty insurer into a global commercial and specialty insurer, as well as price increases and refusing unprofitable business (DePass, 2000).

The organization is committed to supporting the quality of life for all employees and sponsors a variety of programs. A director of employee work/ life quality and balance is responsible for monitoring employee needs and developing recommendations for addressing those needs. The organization was cited by *Working Mother* magazine as one of the 85 best companies to work for as well as by *Fortune* magazine as 66th of the top 100 firms to work for. On-site day care is provided (and has been visited by the First Lady of the United States), as well as programs to improve health. Offering telecommuting and VO working modes is another aspect of the organization's commitment to creating a productive and supportive environment.

SETTING THE STAGE

The architecturally significant corporate headquarters building in downtown St. Paul was built in 1991 and occupies several city blocks near the Landmark Center and the Ordway Theater. The space is finite, and in 1994, managers were asked to seek other venues for housing employees because of the growing number of employees. Teleworking options for employees to work at home one, two or three days a week were offered and in 1999, 4000 employees were using this model. One thousand virtual office employees are now working out of their homes full time. Any employee can propose working in a part-time telecommuting or full-time virtual office mode to his/her manager and human resources, based on the type of work performed and its requirements.

The organization has been experiencing a high degree of change with many layoffs and reductions in force (RIF) due to restructuring efforts and mergers. The competitiveness of the insurance industry parallels many other financial institutions in the post-1980s evolving deregulated environment. Global expansion and mergers with other insurance and financial institutions are common throughout the insurance industry. The St. Paul Companies is represented worldwide by more than 12,000 employees (St. Paul Companies, 2000).

CASE DESCRIPTION

The risk control division of the organization is charged with preventing losses and accidents through customer education, training and on-site advising for customers in construction, commercial, medical technology, manufacturing and the public sector. The specialists provide education in the proper handling of hazardous and construction equipment, behavioral safety, ergonomics, regulatory compliance, and proactive advice to identify and remove potential hazards. In the case of an accident, the risk control specialists are at the site to assist the claims unit to serve the client. The risk control specialists also work closely with the underwriters in helping the clients with the same goal of fully insuring the clients, with sufficient profitability for The St. Paul Companies from the relationship.

The risk control division vice president, Dan Murphy, saw an opportunity to provide a virtual office (VO) working environment for the risk control specialists who worked primarily on construction sites and with construction company home offices. In the pilot, 12 employees were asked to work out of their homes in a virtual office mode (VO). The 12 US regional offices of the St. Paul Companies had provided office space and secretarial support for the risk control specialists, but they were frequently out of the office on client site visits. For those in the St. Paul headquarters, the working paradigm and expectations for corporate center St. Paul employees has shifted dramatically over the past 20 years. Financial services such as banks and insurance companies, once thought of as venerable, stable lifetime employers, have gone through significant efforts to contain and reduce operating expenses. Mergers and acquisitions have accelerated the rate of change, resulting in reassignments of staff and traditional career paths to maximize efficiencies and improve customer service.

In 1998, The St. Paul Companies merged with USF&G to create the eighth largest property-liability insurance company in the US. This presented a significant change in the risk control division with the addition of more risk control specialists. The culture of the organization is a highly competitive organization constantly seeking ways to maximize results and reduce costs, deploying employees to the most appropriate sites and positions to meet the current needs of the organization. Part of the cost reduction efforts included minimizing office space requirements and encouraging individual employee empowerment through use of technology without the assistance of secretaries. The current cultural norm for communication is through e-mail, intranets and Internet for quick access, consistency and low cost delivery.

The risk control division operates within the unique subcultures of the construction and commercial properties industries. Individuals in these cultures gain credibility through training and on-site experience, and the network of individuals in a region is very small, with most knowing each other from having worked together previously. The culture is changing slightly with more diverse populations and underrepresented populations joining the building industry, but those who join must adapt themselves to the prevailing existing hegemonic culture. Due to the ever-present danger inherent in the work, a high degree of trust must be established between the individuals on the job site. That trust carries over to the manager-employee relationship, with the employee's confidence in the judgement and experience of the supervisor in decision making.

When the initial decision was made to move some of the risk control employees out of the regional offices into their homes, very little was provided in the way of preparation and training other than rudimentary technical instruction on use of e-mail and word processing. There were few precedents in the organization for guidance, and the assumption was the problems would be addressed and resolved when encountered. The emphasis in preparation was on the mechanical, technical and logistical aspects of the virtual office, but not on the emotional/affective issues created through distance and lack of physical face-to-face communication and connections. The emotional issues would surface later. The staff was not as technically adept as hoped, and fundamental computing and training issues needed far more effort than anticipated despite the assistance of a help line.

A position was created and staffed in the corporate office to function as an advocate for the original 12 VO employees in 1994, and continues although the number of VO employees is now almost 200. In addition to being a focal point for all employee questions, problem resolution and concerns, a critical contribution of the advocate is to evaluate and modify existing and new corporate center policies for the unique situation of the VO employee. A recent example was a change in the reporting of travel expenses from indirect billing by a travel agency to direct reporting by the employee, which would have created problems for the VO workers. They often must travel great distances by auto and plane and would quickly exceed their monthly expense limits under the new policy, so adjustments were necessary in the electronic expense reporting system. Another example was the charitable giving campaign materials that directed employees to drop off their contributions in the cafeteria, which didn't work for VO employees. The corporate-wide policies are created for the majority of employees who are officed in a corporate center, not those working in VO mode. The VO advocate does internal education of corporate support areas to increase awareness of the needs of the VO employee.

No formal, overall post-implementation feedback study has been conducted, nor has a formalized ongoing feedback forum been created. However, periodic regional meetings have given the risk control specialists an opportunity to share ideas, problem solutions and tips/issues with peers and field supervisors/managers. The corporate center information systems department also provides ongoing training sessions for VO employees. A project initiation report from the perspective of the information systems support group was distributed for discussion in 1999, and plans were formulated based on the review. The IS group partners with the risk control division by supplying training and hardware, as well as providing help desk assistance for VO employees, ensuring standardization of hardware/software platforms, and upgrading equipment and software to current organizational standards. The technology support staff has identified training as the top issue for VO workers and recommended/provided more types of training with more needed on an ongoing basis. Another critical resource is a stable, reliable Internet service provider and hardware/software platform for VO workers. Creating a technology usage proficiency test before launching employees into a VO mode has also been discussed by the IS group to assist VO workers to be as productive as possible with the least frustration.

CURRENT CHALLENGES/PROBLEMS FACING THE RISK CONTROL DIVISION VO

Paradigm Shifting

Within the St. Paul Companies, more than 4,000 employees work in some form of telework—a few days a week to full time home officing. The choice may be voluntary or, in the case of the 190 risk control employees, working in VO mode is a position requirement. This continues to be a challenging paradigm shift for the recently merged USF&G employees who were used to working out of regional offices and on the road with clients. In 1994, all risk control division employees were sent to offices in their homes, and all subsequent hires have been made with the proviso that the employee work out of a home office.

Challenges in identifying the characteristics and traits of a good candidate for VO continue to be discussed in the corporate center. The perception among the interviewees was that a person must be able to work independently with little over-the-shoulder supervision, be technically adept to effectively use all the tools provided to work at home, be able to dedicate a room in one's home for the office space, and be able to manage time demands well. There are no standard industry models or tests commonly accepted for teleworkers, but experiences of other organizations include more restrictions such as requiring the teleworker to sign a contract prior to starting to work at home. The contract may include specified hours of working, required child care for preschool children, confidentiality when working with sensitive data, agreeing to not use the company's equipment for any other purpose than work-related tasks, and other liability limiting actions.

In-depth telephone and e-mail interviews with 16 virtual office employees and 11 corporate center managers and VO staff members showed a wide variety of home office arrangements, ranging from separate buildings near the house to bedrooms or basement space. Each VO employee is provided with a budget for equipment and furniture and tips for setting up effective home work space. The technical office equipment was determined by the central IS group and each line manager. The typical office equipment setups include a laptop computer with a docking station, a monitor, filing cabinet, desk, chair, phone line(s), pager, cell phone, fax, printer, voice mail, company car/truck and expenses, office supplies, line testers for remote log-in testing, and UPS/ surge protectors. In addition to a cell phone, one phone line for each home office is supported by the organization. Some respondents felt strongly that dedicated fax lines were critical for client convenience, along with a line for Internet access, but the cost/benefit was not clearly determined. A multidisciplinary team is currently evaluating cost-effective faxing options for the VO worker. The choice of CompuServe as a provider for dial-up Internet access was made due to the number of local connections available in the US, but local calls are not guaranteed for every risk control employee depending on where he/she lives. Some employees are also living in a community that is outside of cell phone range necessitating a drive to the nearest hilltop to be in range.

Employee Perceptions of VO

Working inside or outside of a corporate or regional office presents challenges as well as opportunities for the employees. The relationships with peers, corporate office managers, information technology staff, and families are redefined with the VO model. The vast majority of those interviewed emphatically felt that working in VO mode was the most effective for themselves and their families. The St. Paul Companies are committed to having VO employees perceived as no different in importance or contribution to the organization than office-bound employees.

Empathy between the VO workers and the corporate center employees suffers due to lack of experience with, and understanding of, each other's situation and exigencies. While all employees are concerned with using expenses wisely, the corporate center employees are responsible for enforcing consistent deployment of organizational policies. They are pushed to achieve very high levels of work product that often require considerable hours outside of the traditional 8 to 5, M-F model. Inside the corporate center and regional offices, the professionals working for and with VO employees are in constant contact and communications with the other corporate control, support and policymaking functional areas. The corporate center employees are the VO employees' representatives for any discussions of policy and procedure changes and the resulting effect on the VO employees.

The VO workers interviewed did not perceive the reality of the daily work lives of the corporate center employees, and were unable to empathize with the constant meetings and long work hours of the corporate center employee. VO employees did not feel that the difficulties they encountered from life on the road and working in isolation were visible to the corporate center employees. The VO workers felt disconnected from the corporate center reality. Yet, many of the corporate center employees themselves had worked in VO mode in different industries and were very empathetic with the VO working challenges. The VO employee also often works unusual hours in response to client demands on weekends and evenings. This also may be done to facilitate personal flexibility of working hours to accommodate family commitments. Several interviewees suggested that in lieu of 'walking in each other's shoes,' videos could be made showing 'normal' day activities of both VO and corporate center employees, then shared with each other to increase empathy and understanding. Each party needs to understand the constraints under which they all work and the measures by which performance is evaluated.

VO employees often felt closer emotionally to the regional underwriters, claims personnel and clients due to a similarity in work backgrounds and experiences, as well as more frequent daily contacts through visits or phone calls. The VO employees often develop a cognitive community with their clients, underwriters and immediate peers/manager, separate and apart from an identification with the corporate part of the St. Paul Companies. One respondent referred to this relationship as similar to the Stockholm syndrome in which hostages develop a sympathetic identification with their captors.

A frequently cited difficulty of working in the home office was the double- edged sword of setting one's own working hours. While it made it possible to have the flexibility to meet family needs for events and communications, it also made it possible for overwork through lack of setting clear boundaries on work vs. personal time during the day and week. Clients often expect immediate responses to cell phone calls, faxes and e-mails any day of the week, due to accidents or site operation hours. Some VO employees reported feeling driven to use any available time to do work, while the vice president of the risk control group emphatically stated that employees should not be risking burnout by doing this. The improvement in quality of life due to increased family interaction through flexible hours and reduced commuting time/expense requirements was a clear incentive for working in the VO mode. Not having anyone in the regional or home office to whom administrative and paperwork tasks could be delegated sometimes resulted in utilization of family members to replenish office supplies or drop off mailings.

A majority of the interviewees cited the benefit of being able to live in a community of their choice as one of the strongest reasons for recommending VO work mode to their peers. Being able to raise a family in a community of choice, locating for the benefit of the spouse's non-VO career or for care of extended family members were strong motivators for dealing with other inconveniences of telework. One respondent was able to move to a family farm, in anticipation of retirement in the same site. Another person reported that being in the same community and region as a majority of her clients gave her the benefit of knowing more about the working environment and climate-related problems experienced by the clients.

Living in a site of choice does not mean that the risk control specialists do not have to travel. They all travel to client sites via plane and car, and some log up to 6,000 miles per month in transit in company vehicles. Visual site inspections and face-to-face communications are a critical part of client relationships in injury prevention and recovery.

Information Technology

Facilitation of any telework arrangement is dependent on use of electronic communications, personal effectiveness, computer applications for self-sufficiency, and communication devices such as cell phones, faxes and pagers. The level and quality of technology support is crucial to the success of the program. Low levels of computer acumen in the teleworkers are common, coupled with the resistance to being self-sufficient in home offices. Many interviewees complained about the time it took to prepare documents and mailings, as well as the need for typing speed/accuracy and knowledge of office procedures such as setting up filing systems. The IS group has made the decision to support one standard hardware and software platform so that licensing arrangements can leverage volume discounts, as well as reducing the training required for help desk support. The initial standards setting for setup, training, help desk, replacements for breakage, software and hardware upgrade distribution and templates for common documents have undergone evaluation and change since the inception of the program.

Telecommunication connection quality and speed present a serious challenge for the VO workers in using dial-up access via CompuServe. The maximum connection speed reported by interviewees was 24 to 28.8 K despite higher capacity modems and frequent time-outs/line drops from the service provider. This may be due to the CompuServe feature that disconnects the connection to the CompuServe service if inactive for 20 minutes while using the web portal, but will require further investigation. Almost everyone interviewed cited the difficulty in downloading graphical images attached to e-mails from other employees who are not aware of the slowness of connection time experienced by VO employees using CompuServe.

There are critical internally developed systems that the VO workers must use in a real-time, on-line mode such as the expense reporting. The slowness of connection time and frequency of line drops makes this an exasperating experience. However, when corporate systems are designed, the VO workers are not the profile of the typical user so VO needs are not always addressed in the initial software release. Access through the corporate LAN while in a corporate office is very fast in contrast to the VO experience with dial-up connections.

The help desk staffing hours are also established for the needs of the majority of the corporate center users and their normal working hours, which are only part of the hours during the week for the VO employees whose schedules are much more flexible. Reaching a help desk employee during nonstandard workweek hours requires several transfers through pagers. The VO employees uniformly stated that it was frustrating for the help desk person to not be able to see the same screens and response times the VO employee was experiencing when the problem occurred. Equipment repairs often necessitate swapping out the equipment to a service center, thus leaving the VO employee without tools for a period of time waiting for the replacement. Laptops are also prone to being dropped or stolen. Several cell phones have wound up dropped in portable toilet units on job sites, and wisely not retrieved for repairs. Although the purchase price of the equipment is depreciated, a monthly maintenance fee is charged through the IS group for the support.

The IS support group is also responsible for distribution of upgrades of software and hardware through CDs and shipments. Ensuring that all St. Paul employees are on similar platforms reduces support. Backing up of C drives on employee laptops is done passively in the background during connection with the network, but employees report the very long connection time (many hours) necessary for the first time backup. Subsequent backups do not take as long, and the software does not have to restart every time a line drops and loses the on-line connection.

The IS group has identified training of VO employees as the number one issue, and the employees would agree. Those interviewed cited the need for even more basic training than originally provided on virus protection and hardware/software/communications fundamental terms and concepts. More classes on self-conducted troubleshooting skills were also requested frequently, in order to reduce dependence on the help desk. Others suggested establishing a VO electronic bulletin board for exchange of questions and tips to more effective use of hardware and software, with input from the IS help desk professionals.

Training for managers on working with VO employees and the need for new performance benchmarks were cited by the interviewees. Also, training on the redefinition of traditional communication and guidance channels was needed for all VO employees.

The scope of training topics requested by the VO employees spanned more topics than technical in nature. Requests for training on topics as diverse as personal time management and filing systems, creating libraries of templates for commonly used documents and presentations, use of e-fax services to send/receive client communications while on the road, home office organization and document management, e-mail composition, typing skill improvement, and refreshers/advanced features on basic software packages were all suggested. Offering training in distance education mode, via WWW-facilitated software would be ideal for the VO worker to access when time was available to study. The technical competence to participate in Web-facilitated training already exists in the VO employee, and time/location issues are the greatest barriers that can be overcome via Web-based training delivery.

Risk Exposure

Broad industry concerns about teleworking engendering risk for organization and the employees includes the equipment, the data, and the employees. Home safety is stressed in training, and the risk control specialists are unique as teleworkers because they are already experts at spotting potential problems. The ergonomic issues associated with working on computers at home, on the road, and on construction sites are not dictated by corporate policies. However, the risk control employees are trained in the problem-prevention discipline and advise their own customers on the prevention of injuries due to poor equipment setup or usage practices, so they know how to prevent their own injuries. The knowledge of what should, and could, be done for proper ergonomic equipment setup may be mitigated by limited equipment budgets for each employee, as well as limited physical space for offices in homes. The organization provides a budget for furnishing the home office as well as a catalog of furniture from which to choose.

The risk control management VO employees do not have a formal employment contract with the organization, although many recommend having these in place to articulate responsibilities and for liability avoidance (Fletcher, 1996). Contracts address responsibilities for all parties as well as ownership and liability issues. Many contracts stress that teleworking arrangements are at the will of the employer, and not a right of the employee. Adding 'incidental business endorsements' to homeowners' policies is also frequently recommended.

Data and equipment protection policies are set by corporate center IS, and VO workers are provided with electrical surge protectors, as well as line testers, for modem use in hotel rooms. The data protection for paper files relies on the safety in the employee's home, and the data stored on the computers is backed up through uploads to the LAN at corporate center. Precautions against viruses are done through use of virus protection software that runs constantly in the background while connected, but many of the interviews revealed a lack of understanding by the employees about how to use the protection software effectively.

As with any type of portable equipment, losses occur from theft and breakage. The theft can be insured, and a level of acceptable breakage incidents can be determined based on experience. Computers do get dropped and other equipment can be damaged accidentally or stolen, but reasonable precautions must be taken by employees.

The difficulty of retrieving the equipment and data from employees who have left the organization, and terminating all corporate system/LAN access, continues to be a challenge. Writing the costs of the equipment off quickly mitigates that loss, but control over misuse or corruption of the data relies on the trusting relationship between the employee and manager.

Measuring Costs and Benefits of Telework

As more organizations plan and implement telecommuting programs, the questions about the actual costs and benefits continue to grow without clear models and guidance. The general perception is that it is a 'good thing to do' and it 'keeps employees happier,' but when pressed for hard numerical results, little can be provided with a high degree of confidence. Few organizations have a clear understanding what it costs to have an office space for an employee in the corporate center, which is necessary before forming any direct comparison to the employee working from home. The costs of floor space are offset from the corporation to the employee who uses part of his/her home for work, as are many of the administrative support costs (although all employees are expected to manage their own word processing and faxing tasks without secretarial support). While some of the communications equipment in the corporate center can be shared (e.g., fax machines and printers), the home office worker must replicate the same or more in terms of equipment accessibility. Furniture expenses are comparable, as well as computer hardware. Accessibility via pagers, cell phones and dial-up access may incur additional costs for teleworkers too.

There are two perspectives for calculating total cost of working environment: from the individual employee's viewpoint and from the corporation's viewpoint. The individual bears the cost of the commute to and from the workplace (e.g., vehicle use, fuel and parking) and ancillary costs for work-quality clothing and meals while on site. However, the value to the individual of having flexibility in scheduling work hours and being closer to the social fabric of family life during the day may be much higher than any hard dollars spent on lunches and new suits. The individual may also be much less inclined to move to new organizations without the teleworking option, hence reducing recruitment and staffing costs for the organization. The benefit of being physically with the client is intangible, but may be measured in continued or increased business, and for risk control specialists, reduced accident claims.

The issue of measuring employee productivity is also challenging. When in the office, more 'water cooler' talk time is consumed, but fewer minutes are wasted waiting for downloads from the LAN or for help desk professionals to return calls. Teleworkers report their tendency to work many more hours per week than they did when in the office, just because the work is right there waiting for them in their home office. Missing the informal conversations in the office is a downside, however, all reported that they were much more focused in the uninterrupted working time spent in home offices. The difficulty in establishing reasonable and effective benchmarks for work is similar for officebound and teleworkers.

From the corporate perspective, the structure of the overhead charge back system in the accounting system may not vary for employees who are not working in corporate center offices. The cafeteria, carpeting, and artwork still come out of the company's earnings, and whether those costs are spread equally across all employees or not is moot. The corporate center building space is fixed and the capacity, not density, of seating is the limiting issue. Renting more office space will consume hard budget dollars as will wiring rental space for LAN and telephone access. The departmental manager may lobby for reduced overhead charges based on having employees without permanent work stations, but will incur travel expenses for bringing the telecommuting employees together for periodic face-to-face training and regional meetings.

Building a reasonable cost/benefit model that incorporates the hard and soft categories, as well as the different intangibles important to each stake-holder, is critical to selling the concept of teleworking arrangements. One person interviewed said that the only reason she kept working for The St. Paul was that she could control her hours and be close to her family during the week. The increased morale levels were apparent in almost every inter-view, except for the individuals who had recently been moved into VO mode after working in a regional office environment. The adjustment time was not yet sufficient.

CONCLUSION

The St. Paul Companies will continue to have risk control employees work in the VO mode so that clients are more effectively served and the limited—and expensive,—office space is used by employees whose positions preclude telecommuting. The VO employees interviewed said they would recommend the arrangement to their friends and colleagues because of the benefits of flexibility of time management and reduction of commuting time. The lessons learned from the past experiences with telework and home office support are being addressed by the support functions within corporate center, but the development of a model for selecting the ideal employee profile for successful telework is underway. The need for additional training is recognized and training sessions are being developed.

REFERENCES

DePass, D. (2000). The St. Paul operating earnings rise 31 percent, *Minneapolis Star and Tribune*, January 28, D1, D3.

Fletcher, M. (1996). Doing your homework, *Business and Management Practices*, 30 (16), 22.

The St. Paul Companies. (2000). [On-line] URL: www.stpaul.com/html [accessed January 1, 2000].

FURTHER READING

Burstein, D., & Kline, D. (1995). *Road warriors: Dreams and nightmares along the information highway*. New York: Penguin Books.

Crandall, N. F., & Wallace Jr., M. J. (1998). *Work and rewards in the virtual workplace: A 'new deal' for organizations and employees*. New York: AMACOM.

Davidow, W. H., & Malone, M. S. (1992). *The virtual corporation: Structuring and revitalizing the corporation for the 21st century*. New York: HarperCollins.

Groth, L. (1999). *Future organizational design: The scope for the IT-based enterprise*. New York: John Wiley & Sons.

Igbaria, M., & Tan, M. (1998). *The virtual workplace*. Hershey, PA: Idea Group Publishing.

Lipnack, J., & Stamps, J. (1997). *Virtual teams: Reaching across space, time and organizations with technology*. New York: John Wiley & Sons.

Weill, P., & Broadbent, M. (1999). *Leveraging the new infrastructure*. Cambridge, MA: Harvard Business School Publishing.

EMPLOYEE
ISSUES

Chapter X

The Impacts of Telecommuting on Organizations and Individuals: A Review of the Literature

Alain Pinsonneault
McGill University, Canada

Martin Boisvert
S.I.X., Inc., Canada

ABSTRACT

Through a review of the literature, this chapter identifies the impacts of telecommuting on organizations and employees and provides recommendations concerning the management of telecommuting. Key success factors of telecommuting programs, such as choosing the right jobs and employees, managerial attitude and expertise, are identified and discussed. Finally, this chapter present several essential steps that organizations should follow when implementing a telecommuting program.

Several factors have contributed to the emergence of telecommuting. First, numerous companies are trying to lower the costs of office space. Second, faced with increased competition, many companies adopt extended workdays and flexible work schedules to better respond to customer needs and to retain and attract skilled employees. Third, computer and telecommunications technologies are becoming increasingly affordable and cost-effective, which enables a strong penetration of Information Technology (IT) in the organization (Brimesk & Bender, 1995). Telecommuting (also known as telework) has grown from its modest beginnings in the early 1970s to achieve an unprecedented level today. More than half of all North American companies currently allow their employees to telecommute (ThirdAge Media, Inc., 1997). Furthermore, the growth of telecommuting is expected to continue in the future. It is estimated that the number of telecommuters in the world should surpass 108 million by 2002 (Gartner Group, 1997).

The objective of this chapter is to provide a comprehensive assessment of what is known on the impacts of telecommuting on organizations and individuals and to present the main elements for effectively managing telecommuting. The chapter contains five sections. First, telecommuting is defined, its components presented, and various types of telecommuting are discussed. Second, the effects of telecommuting on organizations are presented, and third, its effects on individuals are examined. Fourth, the main components essential to effective management of telecommuting are presented. The chapter ends by discussing the managerial implications of this chapter.

DEFINITION AND COMPONENTS OF TELECOMMUTING

Telecommuting can be defined as a "work arrangement in which employees perform their regular work at a site other than the ordinary workplace, supported by technological connections" (Fitzer, 1997, p.65). It can be performed on a full- or part-time basis (i.e., 1, 2, 3, or 4 days/week) and on a permanent or temporary basis (e.g., for one or two months or for the duration of a specific project). Telecommuting represents an expansion of the places and times considered auspicious for work. Three principal components of telecommuting can be identified (Pinsonneault & Boisvert, 1996): utilization of information technology (IT), link with an organization, and delocalization of work. First, telecommuting depends on the processing, manipulation, and transformation of information. Thus, IT represents one of the major components of telecommuting because it enables workers to be in constant communications with their organization and their colleagues. Second, contrary to independent workers, telecommuters have ties with an organization (Bailyn,

1994). Telecommuting is not limited to permanent workers, as employees working on a contractual basis may also telecommute. However, distinctions exist between the type of telecommuting that is prevalent in each group. Authors generally recognize that telecommuting by contractual employees engenders a greater number of difficulties (Huws, 1984, 1993; Ramsower, 1985). This type of telecommuting usually includes clerical work where employees are remunerated on a piecemeal basis or, less frequently, on an hourly basis. Third, telecommuting is not constrained by time and space (Nilles, 1994; Olson, 1988). The delocalization of work takes four main forms: telecommuting from home (home-based), satellite offices, neighborhood work centers, and mobile work. Home-based telecommuting is usually performed in a dedicated area of the worker's place of residence. Satellite offices take the form of small organizational affiliates generally located in proximity to residential areas where a telecommunications link with headquarters is permanently maintained (Doswell, 1992; Nilles, 1994). Neighborhood work centers are private information centers that possess telecommunication tools and that are generally shared by employees from various enterprises (Olson, 1987a; Di Martino & Wirth, 1990; Nilles, 1994). Finally, mobile work is a form of telecommuting that is not limited to any specific "brick and mortar" location. Mobile work empowers employees with the capacity to perform activities in different places and in an ad hoc fashion (e.g., in a car, plane, or hotel room).

THE IMPACTS OF TELECOMMUTING ON ORGANIZATIONS

Figure 1 lists the potential organizational impacts that have been identified in the literature.

Positive Impacts on Organizations

Overall, it seems that managers are satisfied with telecommuting programs. Solomon and Templer (1993) found that 75% of companies that have implemented telecommuting were satisfied or very satisfied with the experience and only 8% were dissatisfied. As Figure 1 indicates, telecommuting has been found to reduce absenteeism and increase employee loyalty to the organization. Olson (1987b), in a study of 20 employees across 20 organizations that had either adopted telecommuting or had started pilot projects, found that telecommuting reinforced the existing relationship between work-

Figure 1: Potential impacts of telecommuting on organizations

Positive Impacts

Lower absenteeism (Duxbury & Higgins, 1995; Fitzer, 1997; Greengard, 1995; Gordon & Kelly, 1986; Huws, 1993; Kraut, 1987; Mahfood 1992; Nilles, 1994; Wilkes, Frolick, & Urwiler, 1994).

Increased feelings of belonging with the organization (Chapman, Sheehy, Heywood, Dooley, & Collins, 1995).

Increased loyalty (Caudron, 1992; Pratt, 1984).

Increased ability to retain best employees and attract new employees (Baig, 1995; Christensen, 1992; Cross & Raizman, 1986; Davenport & Pearlson, 1998; Di Martino & Wirth, 1990; Froggatt, 1998; Gordon & Kelly, 1986; Kraut, 1987; Mahfood, 1992; Olson, 1987b, 1988; Piskurich, 1996; Ruppel & Harrington, 1995).

Increased productivity (Alvi & McIntyre, 1993; Barthel, 1995; Baruch & Nicholson, 1997; Coté-O'Hara, 1993; Duxbury & Higgins, 1995; Gordon & Kelly, 1986; Huws, 1993; Katz, 1987; Kirkley, 1994; Kraut, 1987; Mahfood, 1992; Nilles, 1994; Trembly, 1998; Weiss, 1994; Xenakis, 1997).

Decreased office rental costs and the crowding of offices (Apgar, 1998; Christensen, 1992; Davenport & Pearlson, 1998; Gordon & Kelly 1986; Katz, 1987; Kirkley, 1994; Kraut, 1987; Mahfood, 1992; McCune, 1998; Nilles, 1994, Olson, 1987b).

Quicker responsiveness to customers and to unexpected events (Eldib & Minoli, 1995; Fitzer, 1997; Katz 1987; Korzeniowski, 1997; Nilles 1994).

Increased organizational flexibility (Nilles, 1994; Olson 1987b; Ruppel & Harrington, 1995).

Better usage of information systems (Gordon & Kelly 1986, Hamilton 1987).

Negative Impacts

Absence of best employees (Gordon & Kelly, 1986; Johnson, 1997).

Loss of synergy in the organization (Fitzer, 1997; Hamilton, 1987).

Difficulty in managing telecommuting, which makes supervisors dissatisfied (Christensen, 1992; Fitzer, 1997; Gordon & Kelly, 1986; Katz, 1987; Nilles, 1994; Solomon & Templer, 1993).

Increased data security concerns (Gray, Hodson, & Gordon, 1993; Katz 1987).

Difficulty in objectively evaluating the financial benefits of telecommuting (Alvi & McIntyre, 1993, Doswell, 1992).

ers and their organizations. Moreover, telecommuting was found to allow organizations to retain employees that might otherwise have left and to attract skilled employees who were unwilling to relocate and for whom flexibility was important (Davenport & Pearlson, 1998; Di Martino & Wirth, 1990; Piskurich, 1996).

Second, improved productivity and quality of work associated with telecommuting is probably the most cited organizational benefit in the literature. Some telecommuting specialists evaluate the increase in productivity to be between 15 and 50% (Alvi & McIntyre, 1993; Barthel, 1995; Baruch & Nicholson, 1997; Coté-O'Hara, 1993; Gordon & Kelly, 1986; Kirkley, 1994; Langhoff, 1996; Weiss, 1994). For instance, Huws (1993) reports that telecommuters were 47% more productive (based on evaluations from their supervisors) than their colleagues working in the office. Furthermore, this study indicates that 25% of the work performed by telecommuters was of higher quality than comparable work performed by "traditional" workers. Similarly, New York Telephone reports a 43% average increase in productivity associated with telecommuting, while Control Data Corporation estimates the productivity gain at about 20% (Clutterbuck, 1985). A survey conducted at Northern Telecom (Nortel) indicates that 88% of its telecommuters reported increased productivity ranging from 10% to 22% (Froggatt, 1998). Baruch and Nicholson (1997) found that over 70% of the 62 telecommuters (managers and professionals) they studied perceived themselves as working more effectively than in a traditional work arrangement. Several factors can explain the increase in productivity of telecommuters: lower levels of interference and interruptions, a work environment better tailored to specific individual and task needs, the possibility of choosing more convenient working hours, less time wasted commuting (Eldid & Minoli, 1995), and a stronger focus on achieving the required results rather than simply being physically present at work (Guimaraes & Dallow, 1999). It is important to note, however, that few empirical studies directly and objectively measured productivity gains (Connelly, 1995; Duxbury & Higgins, 1995; Mogelonsky, 1995).

Third, telecommuting allows organizations to reduce certain expenses. Typically, lower costs can be realized from reducing office space, energy consumption, parking spaces, and overcrowding of offices. For example, in the US, IBM reported saving US$75 million by selling buildings and reducing its leased office space (McCune, 1998). Ernst and Young was able to save US$25 million annually by reducing office space by two million square feet (Monnette, 1998). At AT&T, the alternative work initiative is estimated to have saved the company US$460,000 annually (Apgar, 1998).

Fourth, telecommuting allows greater organizational flexibility and a better capacity to quickly respond to unexpected events. In fact, several organizations use telecommuting to decentralize their operations and to organize them into networks. Among the several benefits identified in a survey of 252 IS department heads (Ruppel & Harrington, 1995) was the organization's ability to continue operating in emergency situations (Fitzer, 1997). For instance, following a series of earthquakes, many Californian companies relied on telecommuting to continue their daily operations, and have made these work arrangements permanent due to their initial success (Eldib & Minoli, 1995; Fitzer, 1997). Organizations may also increase their flexibility by hiring workers under various contractual arrangements (e.g., on a temporary basis). Telecommuting also enables an organization to provide flexible working hours for its employees. Finally, telecommuting allows for a more efficient usage of the organization's information system, particularly during non-office hours (e.g., at night and on weekends) (Gordon & Kelly, 1986; Hamilton, 1987).

Negative Impacts on Organizations

Telecommuting can also have negative impacts on organizations. Often, the employees who are better suited for telecommuting (i.e., motivated, well-organized, and requiring little supervision) are those that companies would rather retain on-site (Johnson, 1997). In addition, telecommuting can reduce organizational synergy. Coordination and motivation of employees, valorization of a common culture, and feelings of belonging are much more difficult to sustain in a telecommuting context (Davenport & Pearlson, 1998). In a study aimed at identifying the factors underlying corporate resistance to telecommuting in the UK, Gray, Hodson, and Gordon (1993) found that over 35% of the 115 senior personnel managers surveyed believed that telecommuting threatened corporate structure and identity.

A second negative impact is the discontentment of managers in charge of telecommuters. This often arises due to managers' difficulty in adapting their management styles to the new reality imposed by telecommuting (Christensen, 1992). Third, the security of transmitting corporate data via telecommunication networks concerns managers (Gray, Hodson, and Gordon, 1993; Greenstein, 1999; Katz, 1987). Finally, it is often difficult to objectively evaluate the financial benefits of telecommuting programs (Alvi & McIntyre, 1993; Doswell, 1992).

THE IMPACTS OF TELECOMMUTING ON INDIVIDUALS

Figure 2 presents the potential impacts of telecommuting on individuals.

Positive Impacts on Individuals

As indicated in Figure 2, telecommuting can have both positive and negative impacts on individuals. One of the major benefits of telecommuting stems from the reduction or elimination of the time needed to physically commute to work. Baruch and Nicholson (1997) note that 75.5% of the 62 part-time telecommuters (managers and professionals) they interviewed reported that telecommuting allowed them to save over one hour a day in commuting. By reducing the time and distance that must be traveled to work or by completely avoiding traditional forms of commuting, telecommuting can help alleviate high levels of stress. Moreover, telecommuting provides individuals with more freedom in managing their time between work, leisure, and family responsibilities (Reinsch, 1997). In a survey of 20 part-time telecommuters, increased flexibility and personal control over work and life were judged as being the most important motivators for telecommuting (Knight & Westbrook, 1999). In fact, increased flexibility seems to be associated with higher levels of employee satisfaction and productivity. Ninety percent of Nortel's telecommuters reported greater job satisfaction

Figure 2: Potential Impacts of Telecommuting on Individuals

Positive Impacts

Reduction/ elimination of commute time (Baruch & Nicholson, 1997; Christensen, 1992; DeSanctis, 1984; Mahfood, 1992; Nilles, 1994; Olson, 1988; Pratt, 1984).

Reduced work-related expenses (Baruch & Nicholson, 1997; DeSanctis, 1984; Olson, 1988).

Increased flexibility in the work hours (DeSanctis, 1984; Di Martino & Wirth, 1990; Olson, 1988; Reinsch, 1997).

Increased productivity (Baruch & Nicholson, 1997; Caudron, 1992; Côté-O'Hara, 1993 ; Di Martino & Wirth, 1990; Duxbury & Higgins, 1995; Huws, 1993; Olson, 1988; Reinsch, 1997).

Negative Impacts

Feeling of isolation (Chapman, Sheehy, Heywood, Dooley, & Collins, 1995; Fitzer, 1997; Guimaraes & Dallow, 1999; Huws, 1984, 1993; Johnson, 1997; Katz, 1987; Kinsman, 1987; Olson, 1988; Reinsch, 1997; Solomon & Templer, 1993).

Reduction in chances of promotion (Chapman, Sheehy, Heywood, Dooley, & Collins, 1995; DeSanctis, 1984; Gordon & Kelly, 1986; Hamilton, 1987; Katz, 1987).

Tendency to overwork (Fitzer, 1997; Gordon & Kelly, 1986; Nilles, 1994; Olson, 1988).

Reduction of intra-organizational communication (Ramsower, 1985; Richter & Meshulam, 1993).

---

and 73% reported experiencing a decrease in stress level (McCune, 1998). Richter and Meshulam (1993) note that telecommuting allows individuals to be more productive because it enables them to work when their creativity is high, which for some employees may not correspond with regular office hours. Also, telecommuting lowers the number of distractions in the traditional office environment, thus allowing greater concentration and focus on tasks directly related to one's job (Côté-O'Hara, 1993; Huws, 1993; Nilles, 1994; Pratt, 1984). The results of a survey of 20 telecommuters indicate that 58% of them believed that greater tranquility allowed them to be more productive (Duxbury & Higgins, 1995). Similarly, both Côté-O'Hara (1993) and Di Martino and Wirth (1990) suggest that it is the lower number of interruptions and the increase in attention and motivation due to the elimination of the stress caused by daily commuting that lead to productivity gains. Telecommuting can also lower the costs associated with work, such as travel, clothing and restaurants (Nilles, 1994; Pratt, 1984).

Negative Impacts on Individuals

Telecommuting may also produce some undesirable effects on individuals. Feelings of isolation and the loss of morale are the most commonly cited drawbacks of telecommuting (Fitzer, 1997; Haddon & Lewis, 1994; Huws, 1993; Solomon & Templer, 1993). Almost one third of the 103 telecommuters studied by Reinsch (1997) indicated that being left out of office communications and the feeling of isolation were important disadvantages of telecommuting. Also, Duxbury and Higgins (1995) reported that 23% of supervisors (N = 17) believed that communication problems had arisen during a pilot project conducted at Statistics Canada. However, very little empirical evidence exists to substantiate these claims.

In fact, Katz (1987) argues that lower morale might happen during the first weeks of telecommuting because individuals can feel isolated at first. Apgar (1998) describes one AmEx unit using a 'buddy system' where remote workers contact on-site colleagues each morning. Their topics of conversation are not restricted; instead, the company wishes to promote informal chats about a variety of topics that employees used to engage in "around the water cooler in a commercial office" (p. 135). Companies can stimulate social interactions between co-workers and telecommuters and reduce feelings of isolation by limiting the number of days employees telecommute to less than five and making sure that telecommuters attend company meetings and social events (Guimaraes & Dallow, 1999; Johnson, 1997).

Another potential negative impact related to telecommuting is the emergence of conflicts between family and work related roles (Hartman, Stoner, & Arora, 1991; Richter & Meshulam, 1993). Such situations usually arise when telecommuters, working from their homes, become less productive because distractions and interruptions abound (Mogelonsky, 1995). Results from a survey conducted with 97 telecommuters showed that family disruptions were significantly negatively correlated with telecommuting satisfaction (p = 0.004) (Hartman, Stoner, & Arora, 1991). However, such problems can generally be avoided by ensuring that one room in the home is reserved specifically for telecommuting purposes and that family members support the initiative (Baruch & Nicholson, 1997; Nilles, 1994; Weiss, 1994).

In addition, telecommuters often feel a lack of technical and social support, which is often associated with other problems. A survey of 119 employees in 25 North American companies indicates that lack of organizational support for telecommuters was positively correlated with perceived stress levels (p < 0.001), which was significantly and negatively associated with satisfaction with both the job and non-work life dimensions (Dixon & Webster, 1998).

Some studies indicate that telecommuting on a part-time basis reduces or eliminates several disadvantages (Huws, 1993; Duxbury & Higgins, 1995). Weiss (1994) notes that limiting telecommuting to two or three days a week ensures business and social interactions. A six-month study of 16 telecommuters conducted by Ramsower (1985) indicates that part-time telecommuting arrangements had no effects on the quality and quantity of intra-organizational communications. Conversely, full-time telecommuters communicated with others less often and indicated that it was difficult to ask questions and talk to colleagues. However, Duxbury and Neufeld's (1999) six-month longitudinal study of 36 part-time telecommuters (averaging 15.8 hours/week, standard deviation = 14.82 hours/week) indicates that communications between telecommuters, their managers, and their co-workers were not affected by telecommuting. Similarly, Fritz, Narasimhan, and Rhee (1998) indicate that none of the 170 telecommuters they studied perceived that telecommuting hindered their communications with co-workers.

MANAGING TELECOMMUTING

Because employees are not physically present at the work place, telecommuting often requires a transition from activity- and time-based management to project management and result-based evaluations (Di Martino & Wirth, 1990; Grensing-Pophal, 1999; Guimaraes & Dallow, 1999). Employee evaluation should be conducted in terms of quality, quantity, timeliness, and the degree to which objectives are met, rather than on an hourly basis (Alvi & McIntyre, 1993; Kepczyk, 1998; Johnson, 1997).

Guimaraes and Dallow (1999) studied 316 telecommuting employees in 18 companies and found that, in general, employees perceived their supervisors to be results-oriented rather than activity- or effort-oriented and to have a positive attitude about telecommuting. The telecommuting literature converges with this finding, as it indicates that managers' resistance to telecommuting greatly hinders its diffusion in organizations (Christensen, 1990, Christensen, 1992; Eldib & Minoli, 1995; Haddon & Silverstone, 1992; Huws, 1993; Grensing-Pophal, 1999; Kavan & Saunders, 1998; Solomon & Templer, 1993). A survey of 91 chief executive officers of Canadian companies identified "supervisor attitudes" as one of the most frequently cited problems of implementing telecommuting (Solomon & Templer, 1993). Typically, managers expect and fear the arrival of problems related to the supervision and control of telecommuting employees (Caudron, 1992; Haddon & Silverstone, 1992). The results of a 1995 AT&T survey conducted with 200 senior managers from Fortune 1000 companies shows that 63% of them

Figure 3: Some elements of management in the context of telecommuting

Manager/supervisor should...
- Establish a relationship based on confidence with the workers (Caudron, 1992; Christensen, 1992; Guimaraes & Dallow, 1999; Olson, 1987).
- Enable good communication between telecommuters and other employees (Kirvan, 1995; Staples, 1996; Staples, Hulland, & Higgins, 1998; Guimaraes & Dallow, 1999; Mahfood, 1992).
- Jointly establish precise goals and objectives and assure that sufficient resources are available (Davenport & Pearlson, 1998).
- On a regular basis, evaluate and provide feedback (Gray, Hodson, & Gordon, 1993; Weiss, 1994).
- Ensure that telecommuters participate in organizational activities and are kept informed (Apgar, 1998; Guimaraes & Dallow, 1999).
- Consider telecommuters like any other employee (Fitzer, 1997; Knight & Westbrook, 1999).

perceived that reduced control and supervision represented a major drawback of telecommuting (Johnson, 1997). Typically, managers believe that, because telecommuting decreases the frequency of face-to-face contacts, telecommuters often loaf (Fitzer, 1997). It seems that managerial resistance to telecommuting can often be attributed to a strongly held belief that telecommuters cannot be monitored and effectively managed because of the lack of direct supervision (Caudron, 1992).

The above discussion clearly indicates that as workers move away from the office, managers need to change their managerial style (Richter & Meshulam, 1993). Some elements of effective management in the context of telecommuting can be identified in the literature and are presented in Figure 3. It is clear that one of the necessary conditions for the success of any telecommuting initiative is mutual respect, confidence and trust between employees and managers (Caudron, 1992; Chistensen, 1992; Duxbury & Higgins, 1995; Guimaraes & Dallow, 1999; Olson, 1987a). This condition necessitates that both employees and managers mutually acknowledge each other's abilities and aptitudes. Huws (1993) identifies three different management styles that are being practiced by successful telecommuting managers. The "at hands reach" approach (where the work of employees is closely and frequently monitored) appears to be effective when telecommuters possess minimal qualifications and are remunerated on a piecemeal basis. In the "collaboration" style, targets are mutually agreed upon between teleworkers and managers. In this approach, managers frequently meet teleworkers and supplemental team meetings are often scheduled. This style is especially well suited for permanent employees with whom managers have a close relationship and extensive communications. The "relationship of trust" style (a "laissez-faire" approach) generally applies to self-managed senior professionals that have contractual employment arrangements or work autonomously, and are often paid on an hourly basis.

Notwithstanding the approach favored, telecommuting requires well-structured and constant communications between the telecommuter and the manager. In the survey conducted by Guimaraes and Dallow (1999), telecommuters identified the "ability to communicate well with others" as one of the most important manager characteristic. Moreover, several authors have noted that employees often fear that they will miss out on important information and promotions if they are out of their manager's sight (Adams, 1995; Connelly, 1995; Humble et al., 1995; Weiss, 1994). In a study on the relationship between 103 telecommuters and their managers, Reinsch (1997) found that established relationships began to deteriorate after about seven months into the project. The author suggests that during the first few months, a strong sensitivity exists toward telecommuters because of the newness of the arrangement, but that negative consequences emerge after the initial excitement fades. Because the telecommuter is no longer expected to adhere to regular office hours, managers should be more easily accessible; moreover, it becomes essential to adequately plan meetings, communications, and the exchange of documents (Mahfood, 1992). Hartman et al.'s, (1991) study of 97 telecommuters indicates that adequate communications and technical/ emotional support was significantly and positively correlated to their satisfaction with telecommuting ($p < 0.01$).

Another problem often associated with telecommuting is managers' feeling of losing direct control over employees (Huws, 1984; Kinsman, 1987; Olson, 1987a). Nevertheless, it seems that managers can use alternative strategies to compensate for this loss. These practices include such measures as control by electronic means (e.g., verification of access time and consulted files [Olson, 1985]), increased control by the use of formal and rigorous specifications (e.g., targets, detailed procedures and formalization [Huws, 1984; Kinsman, 1987]), and finally, market control mechanisms (e.g., lump-sum payment based on results or on a piecemeal basis [Haddon & Lewis, 1994]). Guthrie and Pick (1998) reported the attitudes of 134 professionals in the Los Angeles area regarding what represented ethical behaviors when managing telecommuters. Results showed that keeping logs of telecommuters' connect time and performing telephone spot checks were not perceived as being unethical control measures. Also, previous firsthand experience with telecommuting is an important asset for a manager. It provides familiarity with the different facets of telecommuting and enables managers to better respond to the expectations of telecommuting employees (Haddon & Lewis, 1994). Alternatively, managers can reduce reliance on control mechanisms and adopt a more appropriate results-oriented approach. When managers use goals and quotas, they should be reasonable and attainable, but nevertheless challenging for the telecommuter (Davenport & Pearlson, 1998; Mahfood, 1992). Gray, Hodson and Gordon (1993) suggest defining 'milestones' to clearly establish what is expected on what date.

MANAGERIAL IMPLICATIONS

Several benefits can be derived from telecommuting, but to achieve them, managers have to proactively manage the potential business problems associated with telecommuting programs. If one decides to implement a telecommuting program, there are several essential steps that should be followed to maximize the chances of success. Figure 4 presents the main success factors that have been identified in the literature.

First, managers should know why telecommuting is being considered and exactly what benefits are expected. This later allows the organization to evaluate the success of the program and decide whether any corrective measures should be taken. A pilot-project may be necessary to allow the organization to experiment with the various human and organizational dimensions of telecommuting (Apgar, 1998; Christensen, 1990; Guimaraes & Dallow, 1999; Niles, 1994). Weiss (1994) notes that the only real way to evaluate whether telecommuting will be successful in an organization is to experiment with it on a trial basis with three or four employees. At Merrill

Figure 4: Organizational considerations for successfully implementing a telecommuting program

---

Success factors include...

- Clearly establishing the program's goals and objectives (Christensen, 1992; Gray, Hodson, & Gordon, 1993).

- Establishing eligibility to the program, its duration, and its selection process (Apgar, 1998; Christensen, 1992; Guimaraes & Dallow, 1999; Weiss, 1994).

- Ensuring that both managers and employees receive proper training (Davenport & Pearlson, 1998; Fritz, Narasimhan, & Rhee, 1998; McCune, 1998; Staples, 1996; Staples, Hulland, & Higgins, 1998).

- Writing up a formal contract specifying the responsibilities of participants and stating the organization's expectations (Gerber, 1995; McCune, 1998).

- Ensuring an adequate physical and technological environment with all the necessary technical support (Davenport & Pearlson, 1998; Fritz, Narasimban, 1998).

- Determining how telecommuters' performance will be assessed (Fitzer, 1997; Hartman, Stoner, & Arora, 1991).

- Preparing a calendar of events that allows all members of the organization to meet on a regular basis (Christensen, 1990, 1992).

- Ensuring data security (Katz, 1987; Weiss, 1994).

Lynch, for instance, candidates are acclimatized in a simulated home office for a period of two weeks (Apgar, 1998). Such measures not only enable employees to assess for themselves whether telecommuting is right for them, but also minimize the company's risk of selecting inappropriate employees that may want to revert back to traditional office work after an important investment in remote-work equipment has been made. It is also important for managers to monitor the pilot project over a period of several months (Gray, Hodson & Gordon, 1993).

Second, managers should make sure that the home office is equipped with ergonomic furniture and computer equipment (McCune, 1998) and that security measures are enforced (Gray, Hodson & Gordon, 1993; Weiss, 1994). Moreover, if the employer supplies the equipment, the company should specify the limits of personal use (Fitzer, 1997). Clearly describing what performance measures will be used for employee compensation and promotion is also important. Because of management's inability to directly supervise its telecommuters in the traditional sense, compensation based on hourly work may not be appropriate. Instead, measures such as the number of projects completed by the telecommuter (McCune, 1998), on-time deliveries, error rates, quality standards, and customer satisfaction (Fitzer, 1997) may be better. According to the results of a study conducted across 11 organizations with 97 telecommuters, the performance evaluation system adopted by the organization was found to be directly related to telecommuters' productivity ($p < 0.05$) and satisfaction ($p < 0.01$) (Hartman, Stoner & Arora, 1991).

Third, managers need to identify tasks that are suitable for telecommuting. Fritz, Narasimhan, and Rhee (1998) found that higher task predictability was positively associated with increased satisfaction with office communications ($p = 0.004$). Baruch and Nicholson (1997) suggest that telecommuting is especially appropriate for two types of jobs: low autonomy level and technologically simple jobs, for which remote control is easy, and highly autonomous jobs and professions where managers can rely on trust. Similarly, Richter and Meshulam (1993) suggest that both very routine and very complex tasks seem especially adequate for telecommuting. According to Guimaraes and Dallow (1999), tasks that are most appropriate for telecommuting have well-defined objectives and require little need for access to equipment, materials, and services that are available only at the central office.

Fourth, research indicates that some individual characteristics might facilitate telecommuting. Figure 5 presents the tasks and individual characteristics most appropriate for telecommuting.

Figure 5: Task and individual characteristics that better suited for telecommuting

---

Task Characteristics

High informational content that can be decomposed into several sub-tasks easily measurable (Fitzer, 1997; Olszewski & Mokhtarian, 1994).

Requires relatively little face-to-face contact (Duxbury & Higgins, 1995; Fitzer, 1997; Weiss, 1994).

Requires extensive periods of mental concentration (Bailyn, 1994; Fitzer, 1997).

Predictable and have well defined objectives (Fritz, Narasimhan, & Rhee, 1998; Guimaraes & Dallow, 1999).

Does not require specific physical location and does not depend on access to special materials, equipment, or services (Duxbury & Higgins, 1995; Fitzer, 1997; Guimaraes & Dallow, 1999).

For which the steps and expected start and end dates for the various parts of the project can be clearly identified (Gray, Hodson, & Gordon, 1993).

That can be accomplished without continual supervision (Baruch & Nicholson, 1997; Kepczyk, 1998).

Includes either very complex components or routine treatment of information (Richter & Meshulam, 1993).

Based on intensive use of information technology (Di Martino & Wirth, 1990; Weiss, 1994).

Can be easily evaluated in terms of quality, quantity, and timeliness (Baruch & Nicholson, 1997; Johnson, 1997; Kepczyk, 1998).

Individual Characteristics

Autonomous, requiring minimal supervision; very motivated and self-disciplined (Baruch & Nicholson, 1997; Davenport & Pearlson, 1998; Fitzer, 1997; Kepczyk, 1998; Weiss, 1994).

Highly skilled and communicates well with others (Baruch & Nicholson, 1997; Guimaraes & Dallow, 1999; Weiss, 1994).

Knows IT and uses it efficiently (Baruch & Nicholson, 1997; Guimaraes & Dallow, 1999; Kepczyk 1998; Staples, Hulland, & Higgins, 1998).

Has social contacts aside from family and those at work. Prefers work with little social interaction Baruch & Nicholson, 1997; Humble et al., 1995; Richter & Meshulam, 1993).

Volunteers for the telecommuting project (Knight & Westbrook, 1999; De Sanctis, 1983; Huws, 1993; Mahfood, 1992; Nilles, 1994).

Has a trusting relationship with his/her manager/supervisor (Baruch & Nicholson, 1997; Weiss, 1994).

Has a positive attitude towards telecommuting (Guimaraes & Dallow, 1999; Johnson, 1997; Solomon & Templer, 1993).

Highly productive and dependable. Loyal to the organization (Guimaraes & Dallow, 1999; Mahfood, 1992).

Does not have a tendency to overwork (Alvi & McIntyre, 1993; Dunkin & Baig, 1995).

Entrepreneurial and likes flexibility (Bailyn, 1994).

Has a positive self-efficacy assessment about his/her ability to perform remote work (Staples, Hulland, & Higgins, 1998).

Research indicates that it is important that candidates be familiar with the organization's culture and that they have worked in the organization long enough to have established contacts. For example, America West Vacations ensures that it knows the work habits of all of its telecommuters and that, in turn, they know its culture by requiring that candidates have more than one-year work experience in the company (Korzeniowski, 1997). Richter and Meshulam (1993) suggest that young and inexperienced employees do not make good candidates because they have not established contacts within the organizations that they can maintain when telecommuting.

Bailyn (1994), Baruch and Nicholson (1997), Guimaraes and Dallow (1999), Kepczyk (1998), and Nilles (1994) suggest that people who work well in a telecommuting context are self-disciplined, flexible, highly skilled, have a positive attitude toward telecommuting, have high levels of productivity, autonomous and moderately people-oriented, prefer working at home, dependable, and avoid overworking. Employees' self-efficacy has also been found to be associated with telecommuting program success. A survey of 376 remotely managed employees (i.e., in a different building, city, state, or country) from 18 North American organizations indicates that self-efficacy judgments were significantly and positively associated with work effectiveness, overall productivity, and job satisfaction (Staples, Hulland, & Higgins, 1998).

Others have tried to identify different personality traits that might negatively affect telecommuting success. Davenport and Pearlson (1998) suggest that satisfaction with remote work arrangements may eventually suffer when office interactions are important to an employee's social life. Baruch and Nicholson (1997) suggest that certain personality types such as extroverts might feel socially deprived. In addition, they note that less autonomous individuals might feel disoriented when confronted with the high discretionary environment of telecommuting.

Fifth, an important but often-overlooked factor in the success of telecommuting programs is training. Reinsch's (1997) study of 103 telecommuters indicates that 48% of the participants had not received any form of training, and for those who had, they described it as brief, incomplete, and primarily focused on using software. Staples, Hulland, and Higgins (1998) found that prior experience/training was positively associated with remote work self-efficacy, suggesting that organizations should implement adequate training programs to educate both employees and managers about telecommuting. Training should be conducted in three ways: employee training, manager training, and team training (Grensing-Pophal, 1999). Team training consists of bringing a telecommuter and his/her manager together to discuss how the new work arrangement will impact their relationship.

A crucial aspect in both employee and manager training is ensuring that good communication skills develop for telephone, video conferencing, and e-mail exchanges to be effective (Humble, Jacob, & Van Sell, 1995; Grensing-Pophal, 1999) and that managers develop or increase their ability to listen and communicate (Staples, Hulland & Higgins, 1998). It is important for managers not to fall into patterns where communications with employees are strictly task-oriented. Rather, communications should be balanced with some form of relationship-building (Davenport & Pearlson, 1998). In fact, technical and emotional communications and support from managers are critical to ensure employee satisfaction (Hartman, Stoner, & Arora, 1991). Moreover, the entire organization's workforce should also be included in the educational program because telecommuting not only affects the telecommuter but also non-telecommuters who risk losing contact with employees who partake in the program (Apgar, 1998; Grensing-Pophal, 1999). Fritz, Narasimhan, and Rhee (1998) conducted a survey of 170 telecommuters and found that adequate training in IT was crucial in maintaining telecommuters' satisfaction with office communication channels ($p = 0.000$). Training should also encompass such activities as dealing with interruptions by family members and friends, communications with co-workers, and safety and security (Caudron, 1992).

Finally, it appears that part-time telecommuting is more successful than full-time programs. Hartman, Stoner, and Arora (1991) found a negative and significant correlation between the proportion of time spent at home telecommuting and perceived productivity ($p = .04$). Their results indicate that employees who spent 50% or less of their work time telecommuting were more productive ($p = .002$) than those who spent more than 50% telecommuting.

CONCLUSION

This chapter reviewed the business and academic literature and highlighted several important impacts of telecommuting, both positive and negative, on the organization and its employees. Several of the negative impacts described herein can be remedied by adopting telecommuting on a part-time basis. Nevertheless, an organization's decision to implement a telecommuting program must be complemented by effective management practices. And yet, managers often experience difficulty in supervising and objectively evaluating the performance of telecommuters. This occurs, at least in part, because managers may fail to recognize that telecommuting requires a change in managerial style from traditional activity and time-based management to one that has a greater focus on results and the building of a relationship with the telecommuter. This chapter highlights several successful managerial practices and important research findings that may help guide managers in developing their abilities to better manage remote-work arrangements.

The characteristics of the tasks and the employees that are better suited for telecommuting are also presented. Organizations should carefully screen participants to telecommuting programs. Managers should ensure that employees are interested and motivated in telecommuting, self-disciplined, comfortable with using IT, and have good personal contacts within the firm. Furthermore, training is often neglected in practice. Training should not only direct telecommuters in the correct use of the equipment, but also educate employees (both telecommuters and office employees) and managers to recognize the importance of maintaining frequent communications.

ACKNOWLEDGEMENT

Financial support for this research was provided by the Social Sciences and Humanities Research Council of Canada. Thanks are extended to Marlei Pozzebon and Daniel Tomiuk for their help in conducting the research.

REFERENCES

Adams, M. (1995). Remote control. *Sales and Marketing Management,* 147(44), 46-48.

Alvi, S., & McIntyre, D. (1993). The open collar worker. *Canadian Business Review,* 20(1), 21-24.

Apgar, M. V. (1998). The alternative workplace: Changing where and how people work. *Harvard Business Review,* 76(3), 121-136.

Baig, E. C. (1995). Welcome to the officeless office: Telecommuting may finally be out of the experimental stage. *Business Week*, June 26, 104-106.

Bailyn, L. (1994). Toward the perfect workplace? The experience of home-based systems developers. In T. J. Allen & M. S. Scott Morton (Eds.),

Information Technology and The Corporation of The 1990s: Research Studies. New York: Oxford University Press.

Barthel, M. (1995). Telecommuting finding a home at banks. *American Banker*, February 1, 16.

Baruch, Y., & Nicholson, N. (1997). Home, sweet work: Requirements for effective home working. *Journal of General Management*, 23(2), 15-30.

Brimesk, T. A., & Bender, D. R. (1995). Making room for the virtual office. *Association Management*, 47(12), 71.

Caudron, S. (1992). Working at home pays off. *Personnel Journal*, November, 40-49.

Chapman, A. J., Sheehy, N. P., Heywood, S., Dooley, B., & Collins, S. C. (1995). The organizational implications of teleworking. In C. L. Cooper & I. T. Robertson (Eds.), *International Review of Industrial and Organizational Psychology*. New York: Wiley.

Christensen, K. E. (1990). Remote control: How to make telecommuting pay off for your company. *PC Computing*, February, 90-94.

Christensen, K. E. (1992). Managing invisible employees: How to meet the telecommuting challenge. *Employment Relations Today*, Summer, 133-143.

Clutterbuck, D. (1985). *New patterns of work*. New York: Aldershot Gower.

Connelly, J. (1995). Let's hear it for the office. *Fortune*, March 6, 221-222.

Côté-O'Hara, J. (1993). Sending them home to work: Telecommuting. *Business Quarterly*, 57 (3), 104-109.

Cross, T. B., & Raizman, M. (1986). *Telecommuting : The future technology of work*. Homewood, IL: Dow Jones-Irwin.

Davenport, T. H., & Pearlson. K. (1998). Two cheers for the virtual office. *Sloan Management Review*, 39(4), 51-65.

DeSanctis, G. (1983). A telecommuting primer. *Datamation*, 29, 214-220.

DeSanctis, G. (1984). Attitudes toward telecommuting: Implications for work at home programs. *Information and Management*, 7(3), 133-139.

Di Martino, V., & Wirth, L. (1990). Le télétravail, un nouveau mode de travail et de vie [Telecommuting, a new way of work and of life]. *Revue Internationale du Travail*, 129(5), 585-611.

Dixon, T. L., & Webster, J. (1998). Family structure and the telecommuter's quality of life. *Journal of End User Computing*, 10(4), 42-49.

Doswell, A. (1992, October). Home alone? - Teleworking. *Management Services*, 18-21.

Dunkin, A., & Baig, E. (1995, April 17). Taking care of business-without leaving the house. *Business Week*, 106-107.

Duxbury, L., & Higgins, C. (1995). *Rapport sommaire sur le projet-pilote de télétravail de Statistiques Canada* [Summary report on Statistics Canada's pilot-project in telecommuting], January, (Document # 75F0008XPF). Statistics Canada.

Duxbury, L., & Neufeld, D. (1999). An empirical evaluation of the impacts of telecommuting on intra-organizational communication. *Journal of Engineering Technology Management,* 16, 1-28.

Eldib, O. E., & Minoli, D. (1995). *Telecommuting.* Boston: Artech House.

Fritz, M. B. W., Narasimhan, S., & Rhee, H. K. (1998). Communication and coordination in the virtual office. *Journal of Management Information Systems,* 14(4), 7-28.

Fitzer, M. M. (1997). Managing from afar: Performance and rewards in a telecommuting environment. *Compensation and Benefits Review,* 29(1), 65-73.

Froggatt, C. C. (1998, Spring). Telework: Whose choice is it anyway? *Facilities Design and Management,* 18-21.

Gartner Group (1997). *Enterprise remote access: Building the extended workplace.* September 29, Stamford, CT,

Gerber, B. (1995). Virtual teams. *Training,* 32(4), 36-40.

Gordon, G. E., & Kelly, M. M. (1986). *Telecommuting: How to make it work for you and your company.* Englewood Cliffs: Prentice Hall.

Gray, M., Hodson, N., & Gordon, G. (1993). *Teleworking explained.* Chichester, England: Wiley.

Greengard, S. (1994). Making the virtual office a reality. *Personnel Journal,* 73(9), 66-79.

Greengard, S. (1995). All the comforts of home, *Personnel Journal,* 74(7), 104-108.

Greenstein, M. (1999). *Electronic commerce: Security, risk management and control.* New York: McGraw-Hill.

Grensing-Pophal, L. (1999). Training supervisors to manage teleworkers, *Human Resources Magazine,* January, 67-72.

Guimaraes, T., & Dallow, P. (1999). Empirically testing the benefits, problems, and success factors for telecommuting programs. *European Journal of Information Systems,* 8(1), 40-54.

Guthrie, R. A., & Pick, J. B. (1998). Work ethic differences between traditional and telework employees. *Journal of End User Computing,* 10(4), 33-41.

Haddon, L., & Lewis, A. (1994). The experience of teleworking: An annotated review. *The International Journal of Human Resource Management,* 5(1), 193-223.

Haddon, L., & Silverstone, R. (1992). Information and communication technologies in the home: The case of teleworking, Falmer Pict Working paper No. 17.

Hamilton, C. A. (1987). Telecommuting, *Personnel Journal,* April, 91-101.

Hartman, R. I., Stoner, C. R., & Arora, R. (1991). An investigation of selected variables affecting telecommuting productivity and satisfaction. *Journal of Business and Psychology,* 6(2), 207-225.

Humble, J. E., Jacobs, S. M., & Van Sell, M. (1995). Benefits of telecommuting for engineers and other high-tech professionals. *Industrial Management,* 37, 15-19.

Huws, U. (1984). *The new homeworkers: New technology and the changing location of white collar work,* London: Low Pay Unit.

Huws, U. (1993). *Teleworking in Britain* (Report No. 18). London: The Employment Department Research Series.

Johnson, M. (1997). *Teleworking...in brief.* Oxford, UK: Butterworth-Heinemann.

Katz, A. I. (1987). The management, control, and evaluation of a telecommuting project: A case study. *Information and Management,* 13, 179-1190.

Kavan, C. B., & Saunders, C. S. (1998). Managers: A key ingredient to alternative work arrangement program success. *Journal of End User Computing,* 10(4), 23-29.

Kepczyk, R. H. (1998, April-June). Evaluating the virtual office, *The Ohio CPA Journal,* 16-17.

Kinsman, F. (1987). *The Telecommuters.* New York: Wiley.

Kirkley, J. (1994). Business week's conference on the virtual office. *Business Week,* September 5, 63-66.

Kirvan, P. (1995). How to manage systems for remote workers. *Communications News,* 33, 67.

Knight, P. J., & Westbrook, J. (1999). Comparing employees in traditional job structures vs. telecommuting jobs using Herzberg's hygienes and motivators. *Engineering Management Journal,* 11(1), 15-20.

Korzeniowski, P. (1997). The telecommuting dilemma. *Business Communications Review,* April, 29-32.

Kraut, R. E. (1987). Predicting the use of technology: The case of telework. In R. E. Kraut (Ed.), *Technology and the Transformation of White-collar Work*. Hillsdale, NJ: Lawrence Erlbaum Associates.

Langhoff, J. (1996). *The Telecommuter's Advisor: Working in the Fast Lane*. Newport, RI: Aegis Publishing Group, Ltd.

Mahfood, P. E. (1992). *Home Work: How to Hire, Manage and Monitor Employees Who Work at Home*. Chicago: Probus Publishing Company.

McCune, J. C. (1998). Telecommuting Revisited, *Management Review, 87(2)*, 10-16.

Mongelonsky, M. (1995). Myths of telecommuting: Telecommuting might seem like a dream job, but here's the reality- it's still work. *American Demographics, 17*, 15-16

Monnette, C. (1998). Virtual-office workers offer increased productivity to corporations. *Computer Dealer News, 14(13)*, 28.

Nilles, J. M. (1994). *Making telecommuting happen*. New York: Van Nostrand Reinhold.

Olson, M. H. (1983). Remote office work: Changing work patterns in space and time. *Communications of the ACM, 26(3)*, 182-187.

Olson, M. H. (1985). The potential of remote work for professionals. In *Office workstations in the home*. National Research Council. Washington, DC: National Academy Press.

Olson, M. H. (1987a). Telework: Practical experience and future prospects. In R. E. Kraut (ed.), *Technology and the Transformation of White Collar Work*. Hillsdale, NJ: Lawrence Erlbaum Associates.

Olson, M. H. (1987b). *An investigation of the impact of remote work environment and supporting technology*. Working paper, Center of Research on Information Systems, New York University.

Olson, M. H. (1988). Organizational barriers to telecommuting. In W. B. Korte , B. Steinle, and S. Robinson (Eds.), *Telework: Present Situation and Future Development of a New Form of Work Organization*. Amsterdam: North-Holland.

Olszewski, P. & Mokhtarian P. (1994). Telecommuting frequency and impacts for state of California employees. *Technological Forecasting and Social Change, 45*, 275-286.

Pinsonneault, A., & Boisvert, M. (1996). Le teletravail: L'organisation de demain? [Telecommuting: The organization of tomorrow?]. *Gestion, 21(2)*, 76-82.

Piskurich, G. M. (1996). Making telecommuting work, *Training and Development, 50(2)*, 20-27.

Pratt, J. (1984). Home teleworking: A study of its pioneers. *Technological Forecasting and Social Change,* 25, 1-14.

Ramsower, R. M. (1985*). Telecommuting : The organizational and behavioral effects of working at home.* Ann Arbor, MI: UMI Research Press.

Reinsch, N. L., Jr. (1997). Relationship between telecommuting workers and their managers: An exploratory study. *The Journal of Business Communications,* 34(4), 343-369.

Richter, J., & Meshulam, I. (1993). Telework at home: The home and the organization perspective. *Human Systems Management,* 12, 193-203

Ruppel, C. P., & Harrington, S. J. (1995). Telework: Innovation where nobody is getting on the bandwagon. *Data Base Advances,* 26(2-3), 87-104.

Solomon, N. A., & Templer, A. J. (1993). Development of non-traditional work sites: The challenge of telecommuting. *Journal of Management Development,* 12(5), 21-32

Staples, D. S. (1996). An investigation of some key information technology-enabled remote management and remote work issues. *Conference Proceedings of the Australasian Conference on Information Systems,* Hobart, Tasmania, Australia: University of Tasmania.

Staples, D. S., Hulland, J. S., & Higgins, C. A. (1998). A self-efficacy theory explanation for the management of remote workers in virtual organizations. *Journal of Computer-Mediated Communication* [On-line], 3 (4). Available: http://www.ascusc.org/jcmc/vol3/issue4/staples.html [2000, January 13].

ThirdAge Media, Inc. (1997). Telecommuting takes off. *ThirdAge* [On-line]. Available: http://www.thirdage.com/news/archive/971205-05.html [2000, January 12].

Tomaskovic-Devey, D., & Risman, B. J. (1993). Telecommuting innovation and organization: A contingency theory of labor process change. *Social Science Quarterly,* 74(2), 367-385.

Trembly, A. C. (1998). Telecommuting productively. *Beyond Computing,* 7(4), 42-44.

Weiss, J. M. (1994). Telecommuting boosts employee output. *HR Magazine,* 39(2), 51-53

Wilkes, R. B., Frolick, M. N., & Urwiler, R. (1994). Critical issues in developing successful telework programs. *Journal of Systems Management,* 45(7), 30-34

Xenakis, J. J. (1997). Workers in the world disperse! *CFO,* 13(10), 79-85.

Chapter XI

Making Remote Workers Effective

D. Sandy Staples
Queen's University, Canada

ABSTRACT

The use of telecommuting is lower than expected because of manager resistance. This chapter describes research conducted to identify two things: (1) the key issues of working and managing remotely, and (2) the activities that employees and managers should do to increase the effectiveness of remote employees. Three major categories of activities were identified. The first deals with the employee's ability to carry out the right tasks and the manager's ability to assess the employee's effectiveness. The second category reflects the essential role information technology plays in enabling remote work. The third category deals with the employee's need for advice and support. Suggestions are provided for how organizations can make the activities more common. If organizations do this, the potential of telecommuting and virtual office arrangements will be more fully reached as remote employees become more effective and resistance to these new forms of working decreases.

Advances in information technology (IT) are making it possible for more people to work remotely from their main office, co-workers, and managers.

As a result, telecommuting and working from a virtual office are becoming more common. In the past decade or so, there has been some research on telecommuting to understand the practice of working remotely from the main office (e.g., Baruch & Nicholson, 1997; Dixon & Webster, 1998; Fritz, Narasimhan & Rhee, 1998; Guthrie & Pick, 1998; Igbaria & Tan, 1998; Kavan & Saunders, 1998; McCloskey, Igbaria & Parasuraman, 1998; Neufeld, 1997; Ruppel & Howard, 1998). These changing work practices are challenging traditional organizational designs and practices.

Managing employees who are located remotely from their manager is a key issue in telecommuting and virtual organizational structures (Beyers, 1995; Kavan & Saunders, 1998; Tapscott & Caston, 1993). Managers' roles are changing as traditional, hierarchical methods become inappropriate largely because distributed working arrangements make observing behavior more difficult or impractical (Grenier & Metes, 1995; Jenner, 1994; Lucas & Baroudi, 1994; Snell, 1994). The fear of lost managerial control is reported to be a significant factor in preventing widespread adoption of telecommuting by many researchers (e.g., DeSanctis, 1984; Duxbury, Higgins & Irving, 1987; Duxbury & Haines, 1991; Goodrich, 1990; Kavan & Saunders, 1998; Phelps, 1985; Risman & Tomaskovic-Devey, 1989; Roderick & Jelley, 1991). Learning how to manage remote employees well and how to make remote employees effective contributors to their organization should reduce this fear.

The study reported in this chapter adds to the understanding of managing and working remotely and provides guidance to organizations on how to do this better. The study was done in two phases. First, the key issues of working and managing remotely were identified, from both a manager's and an employee's perspective. Second, additional information was gathered to examine the effect of potential best practices and abilities on employee effectiveness. Combined, the information suggests important areas on which organizations should focus with regard to resource allocation and training and skill development of their managers and employees.

This chapter is organized as follows: Previous work relevant to remote work and management of these workers is reviewed. Then the methodology and results of the first phase of the study are presented, followed by the methodology and results of the second phase. The last section discusses the implications of the findings of both phases of research and offers suggestions for practitioners on how to improve their remote working environment.

BACKGROUND

In this study, the concept of remote work was defined as employees working in a physically separate location from their managers. The employees' locations can vary considerably from working at another company office or in their homes to working at a customer's location or out of their cars. Employees working at home instead of in an office supplied by their employer are by definition telecommuting; however, telecommuting is just one work arrangement that results in remote work and management. Telecommuting is only a small part of the virtual workplace, where people work together while being physically distant from each other (Jenner, 1994).

Although literature that addresses remote work and management is sparse and largely based on anecdotal evidence, it does contain some suggestions for managing physically separate people. Developing trust and minimal supervision expectations are important because it is very difficult to supervise and control remote employees due to limited face-to-face contact (Duxbury, Higgins & Irving, 1987; Lucas & Baroudi, 1994; Savage, 1988; Snell, 1994). However, trusting employees often goes against a managerial tradition of control, and a tradition that believes control and efficiency are closely linked and that control is necessary for efficiency (Handy, 1995). Gerber (1995) suggests that trust can be generated in two ways: It comes from developing a relationship through communication and informal activities, as well as from performance (i.e., delivering on promises and being competent).

Information technology appears to play a key enabling role for the new forms of organization (Freedman, 1993; Greengard, 1994; Handy, 1995; Lucas & Baroudi, 1994; O'Hara-Devereaux & Johansen, 1994). The technology allows tasks to be distributed in different places and executed at different times while integrating and effectively controlling the whole process (Mowshowitz, 1994). The virtual workplace provides access to information needed to do a job anywhere, anytime, anyplace; and the latest in communication technology is used to accomplish this (Jenner, 1994). The potentially important role of IT suggests that supporting remote employees' need for IT would be beneficial for organizations.

The ability of remote managers to manage the subtle processes of cultural and political change in organizations is another challenge identified by Walsham (1994). Voss (1996) suggests that although IT offers many opportunities for working and managing remotely, technological solutions cannot replace the basic need for leadership and a shared vision. Morgan (1988) suggests that the values and beliefs (i.e., corporate culture) of an organization serve as a guide to help remote employees move in the same direction in a highly decentralized style of management. Maintaining the corporate culture in a remote environment (Greengard, 1994; Illingworth, 1994) and changing the culture to accommodate the new way of working that IT enables (Beyers, 1995; Lucas & Baroudi, 1994; Morgan, 1988) are challenges that remote employees and managers face.

Many researchers suggest that managers of remote employees need to move toward result-oriented assessments (i.e., a focus on what you contributed today, not whether or not you were at your desk) rather than process-orientated assessments (Baruch & Nicholson, 1997; Duxbury, Higgins, & Irving, 1987; Gerber, 1995; Greengard, 1994; Illingworth, 1994; Metzger & Von Glinow, 1988; Mowshowitz, 1994; Oudekerk, 1996). Mowshowitz (1994) suggests that "the essence of virtual organization is the management of goal-oriented activity in a way that is independent of the means for its realization" (p. 270). Management will have to shift from being a passer of information to being a leader or coach (Duxbury, Higgins, & Irving, 1987; Savage, 1988). This involves a shift towards results-based control rather than behavior-based control.

Beyers (1995) suggests that virtual organizations need virtual management, which implies fewer boundaries, more interaction supported electronically, and increasingly direct communication from one to another without the traditional organizational structures (i.e., moving through a structure to do what must be done). Barner (1996) suggests that managers will need to develop specialized communication and planning skills such as the ability to communicate electronically, without the subtle, nonverbal cues that are present in face-to-face communications. Lucas and Baroudi (1994) suggest that these required changes create misgivings about supervision in the typical manager.

In short, the literature suggests that operating in a remote environment will bring different management challenges and perhaps require a different management style. Operating on a basis of trust between the manager and employee will become more important, as will the need to focus more on the results achieved than on the process used. Managers also will face challenges managing the corporate culture and corporate politics that result from remote working arrangements. It is suggested that IT is a key enabler of remote work. In order to provide further data on these issues and potentially identify others, exploratory research was undertaken with remotely managed employees and managers of remote employees.

PHASE 1 — IDENTIFYING KEY ISSUES

Phase 1 Research Methodology

To identify the key issues and best practices of working and managing remotely, focus groups were conducted with both remote employees and managers of remote employees. Five organizations participated in this phase of the study: three were private sector firms and two were public sector organizations. These five organizations had a large variance in the information technology (i.e., ranging from the telephone to groupware and videoconferencing) used to work remotely and manage remotely, which was advantageous for the study. Employees were selected who worked in a location different from that of their managers. Managers were selected who had employees who reported directly to them and worked in a location different from the employees. Almost all of the employees worked in different cities than their managers so that face-to-face inter-action was infrequent.

A total of 104 people participated in 19 focus groups, split fairly evenly between remote managers (58 participants; 56%) and remote employees (46 participants; 44%). Sixty percent (n=63) of the participants worked in Canada, 37% (n=38) worked in the United States, and 3% (n=3) of the participants worked in England. The managers and employees were not paired. The participants carried out a wide range of job functions.

Phase 1 Findings: Key issues regarding working and managing remotely

After brainstorming about remote environment issues, each participant in the focus groups identified the top three issues from their perspective and ranked them in descending order. For presentation purposes, the key issues were then grouped into categories by the author. Fourteen relatively distinct categories of issues were identified (see Table 1). Although it appears that the 14 categories are unique, as the labels suggest, many of the issues within the various categories are interrelated. Specific issues within each category are described below.

Table 1 was sorted based simply on the total frequency of issues identified in each category. Therefore, equal weightings are given to the participants' choices. To reflect the respondents' importance placed on each selection, a weighted value for each issue category was created by assigning a weight of three for a first choice, a weight of two for a second choice, and a weight of one for a third choice. The resulting values were then summed to

Table 1: Summary of key remote management issues identified by all focus group participants

Issue Categories	1st Choice	2nd Choice	3rd Choice	Total Selections	Percent of Total*	Weighted Score	Rank Using Weighted Score
Communications	34	25	15	74	25.4%	167	1st
Performance management	15	15	17	47	16.2%	92	2nd
Information Technology	11	9	9	29	10.0%	60	3rd
Coaching	5	11	13	29	10.0%	50	4th
Isolation	9	7	8	24	8.2%	49	5th
Travel time/Family balance	2	7	5	14	4.8%	25	7th
Teambuilding	1	5	7	13	4.5%	20	8th
Administrative & peer support	4	5	4	13	4.5%	26	6th
New Employees	2	4	5	11	3.8%	19	9th
Remote environment brings different management challenges	4	0	6	10	3.4%	18	10th
Leadership	4	3	1	8	2.7%	19	9th
Differences across sites	3	2	3	8	2.7%	16	11th
Training	5	1	1	7	2.4%	18	10th
Career Development/Recognition	1	3	0	4	1.4%	9	12th

*Percent of the total number of selections made (Total = 291. This is somewhat less than 3 per participant [3x104 = 312] since not all participants identified key 3 issues).

create the score presented in the two rightmost columns in Table 1. Some small changes occurred when the rankings were done on a weighted score. The rank of teambuilding dropped slightly while the importance of training became somewhat higher. Perhaps the most significant change was that the IT category became clearly more important than the coaching category when a weighted score was used.

Table 1 contains the combined responses of managers and employees. Further analysis was done to examine differences between employees' and managers' key issues (see Table 2). The categories are sorted based on the ranking of the weighted score of the employees. The rightmost column shows the ranking of the weighted score for the managers. The specific issues and the differences between the managers' and employees' views are discussed below. Examining these findings can help organizations learn about

Table 2: Ranking of categories split by employees and managers

Issue Categories	Employees			Managers		
	Percent of Total Selections	Weighted Score	Ranking based on Weighted Score	Percent of Total Selections	Weighted Score	Ranking based on Weighted Score
Communications	19.2%	51	1	30.1%	116	1
Information Technology	17.6%	46	2	4.2%	14	7
Isolation	15.2%	39	3	3.0%	10	8
Administrative and peer support	10.4%	26	4	0.0%	0	12
Performance management	9.6%	21	5	21.1%	71	2
Coaching	7.2%	16	6	12.0%	34	3
Training	4.0%	13	7	1.2%	5	10
New Employees	4.0%	10	8	3.6%	9	9
Travel time/ Family balance	3.2%	7	9	6.0%	18	5
Differences across sites	2.4%	6	10	3.0%	10	8
Career Development/ Recognition	2.4%	6	10	0.6%	3	11
Teambuilding	3.2%	5	11	5.4%	15	6
Remote environment brings different management challenges	1.6%	4	12	4.8%	14	7
Leadership	0.0%	0	13	4.8%	19	4

which types of problems are most common and for whom (i.e., the employee or the manager).

The communications category ranked first in terms of weighted score and percent of total issues identified by both employees and managers. Employees were most concerned with managing communication between the manager and employees and peers for sharing ideas and keeping in touch, as well as with developing the communication skills required for effective remote work. Typical communication issues for managers included: how managers could get timely access to employees and vice versa, and timely response to messages; how to keep remote sites in the communication loop and maintain an equal information transfer for remote and nonremote sites; how to be confident that managers will get the required information from employees about critical issues; and, how to replace the nonverbal signals lost with less face-to-face communications. Both managers and employees were concerned with the loss or lack of informal contact, communication required for effective teamwork, and communication with customers.

Information technology issues were ranked number two by the employees; however, IT issues were number seven in the managers' ranking. Most likely this difference is because employees depended more heavily on IT since it typically was they who worked remotely from the larger corporate office. Therefore, employees would feel the lack of IT support more severely than managers in the larger corporate office (although a substantive number of the managers did identify IT issues).

Information technology issues identified by focus group participants included: the lack of IT equipment and support for remote people; the need for more effective IT to support communication efforts; the need for an infrastructure to meet technical and logistical support needs; the need for an ongoing resource commitment to keep equipment and training up to date; and the need for reliable, 24-hour-per-day access to company networks, information, and support people. Information technology clearly was thought to be a key enabler of effective remote management, and many focus groups had suggestions for improvements within their own organizations.

Isolation issues due to working remotely ranked as the third category for employees but the eighth category for managers. Since typically it is the employees who are isolated, it is not surprising that most isolation issues surfaced from the employees' perspective. Typical isolation issues included: how to deal with the potential for isolation so that remote employees don't feel left out but instead feel part of the team; potential problems due to lack of recognition because employees are remote from headquarters and the harmful effect this can have on their careers; perceptions that those close to the manager have distinct advantages via informal communication and from being involved in the politics; and that support for employee development and employee assistance programs often is weak in remote areas.

Employees ranked lack of administrative assistance and peer support fourth; however, managers mentioned no issues in this category. In most cases, employees are isolated from others in the organization and suffer from the lack of administrative and peer support, so it is logical that this problem should be identified by employees and not by managers.

Performance management issues ranked fifth with employees and second among manager respondents. This difference was not surprising given that performance management tasks normally would be thought of as management responsibilities. Typical performance management issues managers identified included: how to gather information remotely to carry out employee appraisals, especially soft information; how to build trust between managers and employees such that managers feel confident about what their employees are doing; accountability issues including how to measure productivity and shift toward a result-based focus; how to identify employees who are struggling and help them; and how to deliver negative messages remotely.

Employees also were concerned about how to remotely build trust and a relationship between managers and employees, about shifting expectations toward a more results-based orientation, and about collecting accurate productivity information for performance reviews. One issue that only remote employees mentioned was the need to define employee and manager roles in a remote setting so that authority is clearly delegated and coaching responsibilities are clear.

Coaching issues were ranked sixth for employees and third for managers. The most common coaching issue for both managers and employees was the limited access employees have to the manager, and vice versa, caused by working remotely. Other coaching issues mentioned by managers included how to give feedback remotely, how to be a resource for employees and customers, and how to model behavior remotely.

Training issues, seventh in the employees' ranking, identified that it often was difficult for remote people to access training programs and that remote employees needed training to manage their time, communicate effectively, and organize their work. Issues in the new employee category dealt with hiring, orientation and training of new employees for remote postings and with initiating new people into the corporate culture.

Travel time and family balance issues was an important category for managers (fifth), while it ranked ninth in the employee responses. Managers typically suggested that a remote manager's personal life is affected by the time required to travel and that this can cause family conflict, stress, and burnout. It also can be difficult to separate personal and business life given portable computers and other remote tools that make it easy to take work home. Employees, in addition to being concerned about the demands created by travel, were concerned about potential family conflict caused by working at home.

The category labeled differences among sites dealt with issues ranging from problems caused by working across different countries or regions with different languages, cultures and laws, to different standards being adopted at different sites, the perception that different organizational cultures likely develop at different sites. The career development/recognition category dealt with remote employees' potential lack of visibility and recognition and the detrimental effects this could have on their careers.

Teambuilding issues ranked eleventh for employees and sixth for managers, which highlights the difficulty of working and communicating with people located in different locations. Typical teambuilding issues included: building relationships remotely and feeling part of a remote team; creating and sustaining a strong team identity; and effectively building teams with minimal face-to-face contact because of the high costs associated with face-to-face meetings. Leadership issues identified by managers—an important issue category for them—dealt with managing rapid change remotely, building a shared vision and moving the group toward it remotely, leading remotely so employees have the right expectations and priorities, and the difficulties of addressing and planning for strategic issues remotely.

After identifying the top three issues, focus group participants brainstormed ideas about how to address the top issues. A long list of ideas was generated, some of which were specific to the work function or organization. However, a number of general actions managers could take and capabilities that employees should have were identified. These are listed in Appendix B and Phase 2 was carried out to assess how these actions and capabilities impact employee effectiveness, and to identify which are most important.

PHASE 2 — IDENTIFYING PRACTICES AND ABILITIES THAT ENHANCE EMPLOYEE EFFECTIVENESS

Suggestions from the focus group participants were used to identify best practices that managers could adopt to enhance remote employees' effectiveness and potential abilities that employees should have to be able to operate effectively in a remote environment. Suggestions were chosen based on two criteria: (1) they had to be relatively general so they applied regardless of the job function the employee performed, and (2) they could be assessed in a questionnaire. Seventeen manager best practice items and 16 employee abilities were identified.

The purpose of the second phase of the study was to examine the impact of these actions and capabilities on employee effectiveness to identify the actions and capabilities that seemed most important. This information could be of potential value to future researchers and practitioners by indicating areas on which they should focus their attention and efforts. Some focus group participants suggested the "right people" were needed to make remote work effective. Therefore, in addition to manager activities and employee abilities, how employees' demographic and situational characteristics influenced their effectiveness also was examined. This section first describes the methodology used, then summarizes the results.

Phase 2 Research Methodology

A questionnaire was used to facilitate collecting information from a large and geographically disperse sample. This section describes the sampling method, construct measures, and analysis methods.

A questionnaire was sent to 1,343 individuals working in 18 North American organizations that (1) employed individuals who worked remotely from their managers and (2) were interested in participating in a study of remote management. Completed questionnaires were received from 631 respondents for an overall response rate of 47%. Use of the procedure suggested by Armstrong and Overton (1977) (i.e., comparing early respondents with late respondents) indicated no significant differences between respondents and nonrespondents on a variety of demographic variables included in the questionnaire. Thus, nonresponse bias did not appear to be a major problem.

Workers were defined as remote or nonremote based on the physical proximity of their offices to their managers' offices. If employees worked in a different building from their managers (which could be across the city, the state, the country, or even the globe), the employees were considered to be remote workers because they worked remotely from their managers.

Of the returned questionnaires, 376 were from remotely managed employees, representing the actual sample of interest for the study reported here.[1] Forty-seven percent of these respondents worked in private sector high technology firms, 22% worked in private sector financial service firms, and the remaining 31% worked in the public sector. Although all employees included in this sample worked remotely, only 17% actually worked at home (the other 83% worked in corporate offices in different buildings from their managers' offices). The median distance between the respondents' offices and their managers' offices was 483 kilometers.

The questionnaire completed by the respondents contained multiple measurement items relating to each construct in the research model. Wherever possible, appropriate scales that had demonstrated good psychometric properties in previous studies were used. However, for the remaining constructs, sets of items were generated based on reviews of previous relevant literature, expert opinion, or from data gathered in the first phase of this research program. Both a pretest and a pilot study were carried out following the guidelines suggested by Dillman (1978). This was done to achieve acceptable levels of measurement reliability and validity. Questionnaire pretesting was first completed using faculty, graduate student, and practitioner input. This information was used to refine the original survey instrument. A preliminary pilot study questionnaire was then administered to remote employees in one insurance firm, resulting in 64 responses. The resulting data were analyzed and used to further modify the questionnaire items for the full study.

Five measures were chosen to assess various aspects of employee effectiveness. These measures were chosen because they could potentially help employees be effective in an organization. Therefore, these variables had to be relevant and important to organizations and management. One attitudinal variable (organizational commitment) and four behavioral variables (two measures of perceived performance, ability to cope, job stress) were used. Support for choosing these variables follows. Organizational commitment has been found to be negatively related to withdrawal behavior, intention to search for job alternatives, and intention to quit (Mathieu & Zajac, 1990) and burnout (King & Sethi, 1997). Ability to cope is important for making individuals effective in today's fast paced world (Silberman, 1996; Stewart, 1996). High job stress increases organizational costs through increased absenteeism and physical and mental health problems (Bosma, Peter & Marmot, 1998; Shigemi et al., 1997) and has been found to be positively associated with a propensity to leave the job (Rahim & Psenicka, 1996). Performance was assessed by asking respondents their beliefs about the effectiveness of remote work in general as well as about their own overall perceived productivity.

Appendix A contains a list of the items used to measure the constructs and Cronbach's alpha values. All five constructs had acceptable internal consistency. Factor analysis of the 25 items indicated adequate discriminant validity between the constructs. Mean scores were created for each construct and used in the analysis described below. As previously discussed, suggestions from the Phase 1 focus group participants were used to identify best practices that managers could adopt to enhance remote employees' effectiveness and potential abilities that employees should have to be able to operate effectively in a remote environment. Seventeen manager best practice items and 16 employee abilities were identified. These are listed in Appendix B. Semi-structured interviews with six remote employees confirmed this was a relevant, valid list of manager best practices and employee abilities.

Analysis Method

An exhaustive Chi-square-Automatic-Interaction-Detection (CHAID) algorithm (Biggs, de Ville & Suen, 1991) was used to analyze the data. This algorithm is part of SPSS's AnswerTree 2.0 statistical analysis package. Exhaustive search CHAID is a technique for examining a series of predictor variables, adjusting the analysis for the number of predictor variables, and selecting the ones that are statistically significant in predicting a target variable. In this way, meaningful segments, patterns and factors can be identified. For more information on this technique, see the SPSS Website (http://www.spss.com/software/spss/answertree/).

Each of the five effectiveness measures were target variables for this analysis. Three groups of predictor variables were analyzed. The first group of items assessed manager activities. The second group of items assessed employee abilities. The third group comprised respondents' demographic and situational variables. Variables in the third group were education level, tenure in the organization, tenure in the current job, time worked for present manager, distance between respondents' offices and their managers' offices, gender, age, length of time the respondent had been remotely managed, and frequency of face-to-face meetings with their managers.

CHAID analysis produces a tree structure of predictor variables leading to a target variable. The predictor variable in the first level can be interpreted as having the most predictive power for impacting the target variable. In this way, the analysis was used to identify those predictor variables that were most important in terms of affecting employee effectiveness (the target variables in this study).

Phase 2 Results

Results of the CHAID analyses are presented in Tables 3 through 5. The left column in these tables contains the dependent, or target, variable for the specific analysis. These were the five indicators of employee effectiveness: perceived productivity, organizational commitment, job stress, remote work effectiveness perceptions and ability to cope. A separate CHAID analysis was run for each dependent variable; the outcome of the resulting tree is shown in the columns moving from left to right. The column labeled "1st level" reports the predictor variable that was found to be the strongest predictor in the tree analysis. The variables at each level of the tree are split into two or more groups, then the analysis continues identifying which, if any, variable can improve the predictive power of each branch. By examining which variables are significant predictors, and at what level, the importance of the predictor variables on the dependent variables can be determined. The findings of the three groups of predictor variables are reported separately below, starting with manager activities.

Impact of manager activities on employee effectiveness

The ability to predict effectiveness measures from the 17 managerial activities is summarized in Table 3. The results suggest the most important activity is assessing performance based on results because it was the strongest predictor (i.e., at the first level) of the target variable for three of the five effectiveness variables. Related to assessing performance based on results is item 12, communicating goals and setting priorities, as these activities would facilitate assessing results. This activity (i.e., communicating goals and setting priorities) was found to be a significant third level predictor for remote work effectiveness and a significant second level predictor for organizational commitment and job stress. These results further support the importance of being able to assess performance based on results.

Support for information technology (IT) needs was the strongest predictor of job stress. It also was a second-level predictor for organizational commitment and remote work effectiveness, while encouragement to use IT

Table 3: The ability to predict effectiveness outcomes by manager activities

Target Variable	Significant Predictors - Manager Activities			
	1st level	2nd level	3rd level	4th level
Productivity	16. Assesses my performance based on the results I achieve rather than how I spend my time	10. Keeps his/her voice message updated so callers know when they can expect a response	13. Is available for consultation and advice	7. Encourages me to use available information technology tools effectively
Organizational Commitment	16. Assesses my performance based on the results I achieve rather than how I spend my time	15. Supports my information technology needs with equipment, financial support and training	14. Supports and promotes social activities and team-building activities	12. Communicates goals and sets priorities with me
		8. Sets expectations about the frequency, method, and subjects of communications between the two of us	17. Keeps me up to date on news within the company	
Job Stress	15. Supports my information technology needs with equipment, financial support and training	13. Is available for consultation and advice	6. Uses and runs teleconference calls effectively	12. Communicates goals and sets priorities with me
Remote Work Effectiveness	13. Is available for consultation and advice	15. Supports my information technology needs with equipment, financial support and training	12. Communicates goals and sets priorities with me	
		9. Takes the time for informal discussions with me so that a relationship develops between us		
Ability to Cope	16. Assesses my performance based on the results I achieve rather than how I spend my time	11. Keeps an accessible schedule so that people know where to locate him/her		

was a fourth-level predictor of productivity. This pattern suggests that providing IT support can be the second-most important manager activity from among the list tested here.

Being available to provide consultation and advice was the level-one predictor for remote work effectiveness perceptions, a second-level predictor of job stress, and a third-level predictor of perceived productivity. Keeping an updated voicemail message and an accessible schedule, second-level predictors of productivity and ability to cope, respectively, also are related to the idea of being accessible. Taking time for informal discussions, an activity related to providing advice, was found to be a significant level-two predictor of remote work effectiveness perceptions. Therefore, the results suggest that being available is a key activity for managers of remote employees.

The remaining significant predictors deal with communications and social/team building activities. These were all significant predictors of organizational commitment only, indicating that these activities, although important, have less effect on other aspects of effectiveness.

Impact of employee abilities on employee effectiveness

Table 4 summarizes the five CHAID analyses that were done to identify which employee abilities appeared to best predict the five effectiveness measures. The results suggest that employees' abilities to set objectives that align with their managers' is perhaps the most important ability. It was a level-one predictor for job stress, ability to cope and perceptions of remote work effectiveness, and a second-level predictor for organizational commitment. Related to the ability to set objectives is the ability to achieve them. This was reflected in two items: the ability to prioritize tasks to use time effectively and the ability to complete the priority tasks. Both of these items were significant predictors of perceived productivity. Collectively these three abilities could be thought of as having the ability to make effective use of time (i.e., do the right thing).

The second most common group of significant predictors deals with employees' abilities to get help and information. Being able to contact the manager immediately was the first-level predictor of organizational commitment and a second-level predictor of job stress. Being able to access the manager within the same day was a third-level predictor of productivity and job stress. The fact that item 5, being able to locate a manager within two to three days, was not a significant predictor suggests that being able to contact the manager in a shorter time than that is more important. Being able to access support staff, knowing which co-workers need information, and being able to access the information needed to carry out their jobs were all significant predictors for most of the measures of effectiveness. These findings support the need for remote employees to have ready access to their managers as well as to co-workers who have knowledge important to the employees' ability to carry out their tasks.

The last two groups of employee capabilities deal with using information technology and being able to organize their own offices. The ability to learn how to use information technology was a second-level predictor of remote

Table 4: The ability to predict effectiveness outcomes by employee capabilities

Target Variable	Significant Predictors — Employee Capabilities				
	1st level	2nd level	3rd level	4th level	5th level
Productivity	2. Prioritize tasks to use my time effectively	3. Complete my daily priority tasks	4. Get a response from manager to a request for help within the same day 9. Access appropriate support staff readily	16. Access information needed to perform my job in an efficient manner	
Organizational Commitment	6. Locate my manager & contact him/her immediately	7. Set objectives that align with my manager's goals 8. Know which co-workers to go to for specific information	9. Access appropriate support staff readily		
Job Stress	7. Set objectives that align with my manager's goals	6. Locate my manager and contact him/her immediately 15. Set up a filing system to organize work documents	4. Get manager's response within the same day on a request for advice	12. Learn a new software package when an instructor is present to guide me 6. Locate my manager and contact him/her immediately	14. Organize office equipment, desk & papers effectively
Remote Work Effectiveness	7. Set objectives that align with my manager's goals	11. Learn a new software package via written material 16. Access information needed to perform my job in an efficient manner	3. Complete my daily priority tasks		
Ability to Cope	7. Set objectives that align with my manager's goals	8. Know which co-workers to go to for specific information 14. Organize office equipment, desk and papers effectively	10. Learn how to use a computer via written material 9. Access appropriate support staff readily	11. Learn a new software package via written material	15. Set up a filing system to organize work documents

work effectiveness perceptions, a third- and fourth-level predictor of ability to cope, and a fourth-level predictor of job stress. Being able to organize their own office (i.e., setting up a filing system and generally organizing the office) were second-level and fifth-level predictors of job stress and ability to cope.

Impact of situational variables and demographic characteristics on employee effectiveness

Table 5 summarizes the results of this series of CHAID analyses. The length of time employees had been working remotely from their managers was the most significant demographic variable. It was a first-level predictor in three cases and a second-level predictor in one other case. Frequency of face-to-face meetings was a first-level predictor for ability to cope. The length of time employees had worked for their present managers was a second-level predictor of perceptions of remote work effectiveness. No other demographic variables were significant predictors. The results suggest that effectiveness of remote employees increases as their experience with this way of being managed increases and that the other demographic variables included here are not that important in determining a remote worker's effectiveness.

Table 5: The ability to predict effectiveness outcomes by situational and demographic characteristics of the remote employee

Target Variable	Significant Predictors — Demographic Characteristics of the Remote Employee	
	1st level	2nd level
Productivity	Length of time been remotely managed	
Organizational Commitment	Length of time been remotely managed	
Job Stress	No significant predictors	
Remote Work Effectiveness	Length of time been remotely managed	Time worked for present manager
Ability to Cope	Frequency of face-to-face meetings with manager	Length of time been remotely managed

DISCUSSION OF THE FINDINGS

The results from this study suggest there are three major categories of activities that employees should be able to do and that managers should do to increase the effectiveness of remote employees. The first deals with the employee's ability to carry out the right tasks and the manager's ability to assess how well the employee achieves this. The second category reflects the essential role IT plays in enabling remote work. The third category deals with the employee's need for advice, help and support.

Doing the right thing and being recognized for it

Clearly assessing results based on performance is a critical task for managers to do well. Communicating goals clearly and setting priorities with employees helps make this possible. Participants of the focus groups offered managers other suggestions on how to assess results-based performance. These included setting project milestones; having periodic reviews, established check-in periods, and frequent updates; making sure communications are clear and understood; establishing a regular communication pattern (e.g., chat every two days at a specific time); agreeing on what the results indicators will be and on how to measure and track them; getting regular feedback from the employee's co-workers and customers; and collecting specific examples of performance-related actions and results so that objective data can be used in performance discussions.

Employees have to have the ability to set objectives that match those of their managers. The actions above will facilitate this. Of course, being able to set objectives is not effective unless one is able to reach the objectives. Using time effectively was identified as a key ability for employees to have. This includes being able to prioritize daily tasks and completing the high-priority tasks. This implies time management training and strategies would be very valuable for remote employees. Commercial training courses for these skills are readily available, both in face-to-face formats and on video and audio-tapes. Coaching from managers and co-workers who are proficient in these skills also would be very valuable.

Impediments to efficiency in a remote office are poor office design, equipment, and filing systems. Focus group participants suggested that often it is assumed incorrectly that the remote employee knows how to establish an effective office. Establishing effective filing systems and worrying about the ergonomic designs of office equipment are skills that employees often do not possess. Providing remote employees with this knowledge, either via training or via skilled assistance, can increase their day-to-day effectiveness.

The need for information technology

Supporting employees' need for IT tools and training increases remote employees' effectiveness. Information technology support includes providing equipment, training to use the equipment, and support staff and material to help solve problems. Managers can help by obtaining the support of their peers and supervisors for allocating the required resources. Making IT training an explicit expectation and objective for employees, as well as providing funding and allocating time, would stimulate employees to learn how to use the IT resources at hand.

Since communication in a remote setting often is done via IT, having reliable equipment and network connections are more important to remote workers than nonremote workers. Therefore, if resources are scarce, priority should be given to remote employees. According to the focus group participants, this often is not done. The "best" equipment often is given to those in the head office, since they are more visible and possibly more vocal.

One advantage of remote working is flexible working hours. Since remote employees depend on IT heavily and may be working any hour of the day or week, having IT support staff available who can help with problems and questions 24 hours a day, seven days a week (including visiting the remote employee's location) is a growing need. One employee focus group participant summed up the dependence on information technology by stating "IT is our lifeline." Remote employees must have reliable communication systems and the ability to use them well.

The ability to get help and information when needed

Employees' ability to reach people for help and advice was identified as significantly impacting remote employees' effectiveness. The people they need to reach could include managers, co-workers, or support staff. To assist with this, remote employees should be provided with contact information, such as on-line phone number and e-mail address lists, that is kept up to date. Before employees try to contact co-workers for help, they need to know who has the information they need. A knowledge management system could help, such as a listing of each employee's expertise. A knowledge repository could be even more effective; it could contain lessons learned from various situations that employees could access and use to help resolve current issues. See Alavi and Leidner (1999) for more information on current trends in and benefits of knowledge management systems.

The manager's availability to provide employees with advice and help was identified as significantly impacting remote employees' effectiveness.

Doing things that enable employees to know how and when they can reach the manager, such as having an accessible schedule and a current voicemail message, are actions that facilitate this. Focus group participants identified that setting up a routine process for making contact also was important. For example, a short (e.g., 15-minute) telephone call daily or a few times per week would be a good practice. Setting expectations about when one can expect a message to be returned and then meeting those expectations also is an important practice to follow. The results in this study suggest that being able to reach or get a response from the manager within a day is required. Possibly a longer delay means the help comes too late to be valuable. This suggests that if managers are going to be incommunicado for more than a day, they should clearly designate others who can act in their place.

So what does all this mean for organizations? Length of time being managed remotely was a significant predictor of employee effectiveness, suggesting that employees improve as experience accumulates. Since time can't be compressed, the real question for organizations, then, is how to reduce employees' learning curves. Training and developing employees and managers so they have the abilities to practice the actions identified above is perhaps the most obvious way to do this (other than hiring people who have had previous remote work experience). Organizations have to allocate resources to do this training, as well as resources to support the other activities mentioned (e.g., for IT equipment, support staff, etc.), and they must recognize and reward managers and employees who successfully carry out the required practices. Setting up mentoring programs for inexperienced remote employees to learn from experienced remote employees would help reduce the learning curve. Mentoring programs also would be of value for remote managers. Any mechanisms that allow remote employees and managers to share best practices and support one another would help facilitate overall learning in the organization.

When considering the recommendations in this chapter, it is worth reflecting on the limitations of study. The samples used in this study included respondents from several organizations who did a wide variety of job functions. This is a strength in terms of the ability to generalize the results; however, it can also be considered a limitation since there was no control for potential confounding factors such as job task or organizational culture. Consequently, it must be left to future research work to determine, for example, whether the results are consistent between high technology and non-technology workers. Also, the second phase of the study relied heavily on individual employees' perceptions collected through a large-scale mail survey. Alternative methods of data collection could be used in future studies. For example, assessments of productivity could be obtained from respondents' managers and co-workers, or from more objective sources.

CONCLUSION

Previous research has found the use of telecommuting is lower than expected because of manager resistance. The results presented here should reduce this resistance by educating managers about what they can do to help make their remote employees effective team members. A number of abilities that remote employees should have were identified. Managers can train and coach their remote employees so that employees learn these skills and develop the abilities required to be effective in a remote work environment. Organizations need to support remote managers with training and resources so that a strong IT infrastructure is provided, employees can obtain help and knowledge as required, and both managers and employees know how to set objectives, meet them, and assess the results. If organizations do this, the potential of telecommuting and virtual office arrangements will be more fully reached as remote employees become more effective and resistance to these new forms of working decreases.

REFERENCES

Alavi, M., & Leidner, D. E. (1999). Knowledge management systems: issues, challenges, and benefits. *Communications of the Association for Information Systems,* 1(7), [On-line] Available: http://cais.isworld.org/contents.asp. [September 6, 1999].

Armstrong, J.S., & Overton, T.S. (1977). Estimating non-response bias in mail surveys. *Journal of Marketing Research,* 14(3), 396-402.

Barner, R. (1996). The new millennium workplace: seven changes that will challenge managers and workers. *The Futurist,* 30(2), 14-18.

Baruch, Y., & Nicholson, N. (1997). Home, sweet work: requirements for effective home working. *Journal of General Management,* 23(2), 15-30.

Beyers, M. (1995). Is there a future for management? *Nursing Management,* 26(1), 24-25.

Biggs, D., de Ville, B., & Suen, E. (1991). A method of choosing multiway partitions for classification and decision trees. *Journal of applied statistics,* 18(1), 49-62.

Bosma, H., Peter, R., & Marmot, M. (1998). Two alternative job stress models and the risk of coronary heart disease. *American Journal of Public Health, 88* (1), 68-74.

DeSanctis, G. (1984). Attitudes toward telecommuting: Implications for work-at-home programs. *Information & Management,* 7(3), 133-139.

Dillman, D. A. (1978). *Mail and Telephone Surveys: The Total Design Method.* New York: Wiley.

Dixon, T. L., & Webster, J. (1998). Family structure and the telecommuter's quality of life. *Journal of End User Computing,* 10(4), 42-49.

Duxbury, L., & Haines, G., Jr. (1991). Predicting alternative work arrangements from salient attitudes: A study of decision makers in the public sector. *Journal of Business Research,* 23(1), 83-97.

Duxbury, L. E., Higgins, C. A., & Irving, R. H. (1987). Attitudes of managers and employees to telecommuting. *Infor,* 25(3), 273-285.

Freedman, D. H. (1993). Quick change artists. *CIO,* 6(18), 32-28.

Fritz, M. W., Narasimhan, S., & Rhee, H. (1998). Communication and coordination in the virtual office. *Journal of Management Information Systems,* 14(4), 7-28.

Gerber, B. (1995). Virtual teams. *Training,* 32(4), 36-40.

Goodrich, J. N. (1990). Telecommuting in America. *Business Horizons,* 33(4), 31-37.

Greengard, S. (1994). Making the virtual office a reality. *Personnel Journal,* 73(9), 66-79.

Grenier, R., & Metes, G. (1995). *Going Virtual: Moving your organization into the 21st century.* Upper Saddle River, NJ: Prentice Hall.

Guthrie, R. A., & Pick, J. B. (1998). Work ethic differences between traditional and telework employees. *Journal of End User Computing,* 10(4), 33-41.

Handy, C. (1995). Trust and the virtual organization. *Harvard Business Review,* 73(3), 40-50.

House, R. J., Schuler, R. S., & Levanoni, E. (1983). Role conflict and ambiguity scales: reality or artifacts? *Journal of Applied Psychology,* 68(2), 334 - 337.

Igbaria, M., & Tan, M. (1998). *The Virtual Workplace.* Hershey, PA: Idea Group Publishing.

Illingworth, M. M. (1994). Virtual managers. *InformationWeek,* 479(June 13), 42-58.

Jenner, L. (1994). Are you ready for the virtual workplace? *HR Focus,* 71(7), 15-16.

Kavan, B. C., & Saunders, C. S. (1998). Managers: a key ingredient to alternative work arrangements program success. *Journal of End User Computing,* 10(4), 23-32.

King, R. C., & Sethi, V. (1997). The moderating effect of organizational commitment on burnout in information system professionals, *European Journal of Information Systems,* 6(2), 86-96.

Lucas, H. C., Jr., & Baroudi, J. (1994). The role of information technology in organization design. *Journal of Management Information Systems,* 10(4), 9-23.

Mathieu, J. E., & Zajac, D. (1990). A review and meta-analysis of the antecedents, correlates, and consequences of organizational commitment, *Psychological Bulletin,* 108(2), 171-194.

McCloskey, D. W., Igbaria, M., & Parasuraman, S. (1998). The work experiences of professional men and women who telecommute: convergence or divergence. *Journal of End User Computing,* 10(4), 15-22.

Metzger, R. O., & Von Glinow, M. A. (1988). Off-site workers: at home and abroad. *California Management Review,* 30(3), 101-111.

Morgan, G. (1988). Developing skills in remote management. In *Riding the Waves of Change: Developing Managerial Competencies for a Turbulent World.* San Francisco: Jossey-Bass Publishers.

Mowday, R. T., Steers, R. M., & Porter L. W. (1979). The measurement of organizational commitment, *Journal of Vocational Behavior,* 14(2), 224-247.

Mowshowitz, A. (1994). Virtual organizations: A vision of management in the information age. *The Information Society,* 10(4), 267-288.

Neufeld, D. J. (1997). *Individual Consequences of Telecommuting.* Unpublished Doctoral Thesis. University of Western Ontario, London, Canada.

O'Hara-Devereaux, M., & Johansen, R. (1994). *Globalwork: Bridging distance, culture, and time.* San Francisco: Jossey-Bass Publishers.

Oudekerk, D. (1996). Getting them ready for virtual work. *Training,* 33(6), 14-16.

Phelps, N. (1985). Mountain Bell: Program for managers. In *Office Workstations in the Home.* Washington, DC: National Research Council, National Academy Press.

Rahim, M.A., & Psenicka, C. (1996). A structural equations model of stress, locus of control, social support, psychiatric symptoms, and propensity to leave a job. *The Journal of Social Psychology,* 136(1), 69-79.

Risman, B. J., & Tomaskovic-Devey, D. (1989). The social construction of technology: Microcomputers and the organization of work. *Business*

Horizons, 32(3), 71-75.

Rizzo, J., House, R., & Lirtzman, S. (1970). Role conflict and ambiguity in complex organizations. *Administrative Sciences Quarterly,* 15(2), 150-163.

Roderick, J. C., & Jelley, H. M. (1991). Managerial perceptions of telecommuting in two large metropolitan cities. *Southwest Journal of Business & Economics,* 8(1), 35-41.

Ruppel, C., & Howard, G. S. (1998). The effects of environmental factors on the adoption and diffusion of telework. *Journal of End User Computing,* 10(4), 5-14.

Saks, A. M. (1995). Longitudinal field investigation of the moderating and mediating effects of self-efficacy on the relationship between training and newcomer adjustment. *Journal of Applied Psychology,* 80(2), 211-225.

Savage, J. A. (1988). California smog fuels telecommuting plans. *Computerworld,* 22(18), 65-66.

Shigemi, J., Mino, Y., Tsuda, T., Babazono, A., & Aoyama, H. (1997). The relationship between job stress and mental health at work. *Industrial Health,* 35(1), 29-35.

Silberman, S. J. (1996). Managers' ability to cope with change will make or break their career, *Sales & Marketing Management,* 148(2), 20-21.

Snell, N. W. (1994). Virtual HR: Meeting new world realities. *Compensation & Benefits Review,* 26(6), 35-43.

Stewart, T. A. (1996). Leading edge: Tools that make business better. *Fortune,* 134(12), 237-240.

Tapscott, D., & Caston, A. (1993). *Paradigm Shift: The New Promise of Information Technology.* New York: McGraw-Hill.

Voss, H. (1996). Virtual organizations: The future is now. *Strategy & Leadership,* 24(4), 12-16.

Walsham, G. (1994). Virtual organization: An alternative view. *The Information Society,* 10(4), 289-292.

ENDNOTES

1   The questionnaire collected data that allowed identification of remotely managed respondents. In this way, participating organizations did not have to specifically identify remotely managed respondents, and data was gathered from locally managed employees for purposes not reported here.

APPENDIX A

Items used in the questionnaire to measure
the dependent variables

Item Wording

ABILITY TO COPE (Cronbach's alpha of the 4 items = 0.87) - Source: House, Schuler and Levanoni (1983); Saks, (1995)

I frequently don't know how to handle problems that occur in my job*

I often find that I cannot figure out what should be done to accomplish my work*

I am frequently confused about what I have to do on my job*

I am frequently unsure about how to do my work*

OVERALL PRODUCTIVITY (Cronbach's alpha of the 5 items = 0.89) - Source: expert opinion, developed for this study

I believe I am an effective employee

Among my work group, I would rate my performance in the top quarter

I am happy with the quality of my work output

I work very efficiently

I am a highly productive employee

REMOTE WORK EFFECTIVENESS (Cronbach's alpha of the 4 items = 0.79) - Source: expert opinion, developed for this study

Working remotely is not a productive way to work*

It is difficult to do the job being remotely managed*

Working remotely is an efficient way to work

Working remotely is an effective way to work

ORGANIZATIONAL COMMITMENT (Cronbach's alpha of the 7 items = 0.90) - Source: Mowday, Steers and Porter (1979)

I promote my organization to my friends as a great organization to work for

I would accept almost any type of job assignment in order to keep working for my organization

I find that my values and the organization's are similar

I am proud to tell others that I am part of this organization

My organization really inspires the very best in me in the way of job performance

I am extremely glad that I chose this organization to work for over others I was considering at the time I joined

For me, this is the best of all possible organizations for which to work

JOB STRESS (Cronbach's alpha of the 5 items = 0.84)) - Source: Rizzo, House and Lirtzman (1970)

I work under a great deal of tension*

I have felt fidgety or nervous as a result of my job*

If I had a different job, my health would probably improve*

Problems associated with my job have kept me awake at night*

I often "take my job home with me" in the sense that I think about it when doing other things*

---

* after the item label designates reverse coding. All questions used a 1 to 7 scale, with the anchors being "strongly disagree" and "strongly agree."

APPENDIX B
Manager's activities and employee capability items used in the questionnaire

Item Wording	Item Code
POTENTIAL BEST PRACTICES BY MANAGER (My manager…)	
Runs meetings effectively (e.g., sets agendas, publishes minutes, designates a chairperson)	1
Has good communication skills (e.g., a good listener, picks up on nonverbal cues, asks for clarification when needed, and sets positive tone of discussion)	2
Asks for and listens to my ideas and solutions	3
Uses e-mail effectively to send information updates to the work group	4
Uses available information technology tools effectively	5
Uses and runs teleconference calls effectively (e.g., sets objectives & format, encourages participation)	6
Encourages me to use available information technology tools effectively	7
Sets expectations about the frequency, method, and subjects of communication between the two of us	8
Takes the time for informal discussions with me so that a relationship develops between us	9
Keeps his/her voice message updated so callers know when they can expect a response	10
Keeps an accessible schedule so that people know where to locate him/her	11
Communicates goals and sets priorities with me	12
Is available for consultation and advice	13
Supports and promotes social activities and team building activities	14
Supports my information technology needs with equipment, financial support and training	15
Assesses my performance based on the results I achieve rather than how I spent my time	16
Keeps me up to date on news within the company	17
EMPLOYEE ABILITIES (To aid in performing my job, I could…)	
Set objectives that align with the organization's goals	1
Prioritize tasks to use my time effectively	2
Complete my daily priority tasks	3
Get a response from my manager for a request for advice or help within the same day	4
Get a response from my manager for a request for advice or help within 2 to 3 days	5
Locate my manager and contact him/her immediately	6
Set objectives that align with my manager's goals	7
Know which of my co-workers to go to for specific information	8
Access appropriate support staff readily	9
Learn how to use a computer when I am provided with written instructional material	10
Learn a new software package when I am provided with written instructional material	11
Learn a new software package when an instructor is present to guide me	12
Use a fax machine to send documents	13
Organize my office equipment, desk and papers effectively	14
Set up a filing system to organize work documents	15
Access information needed to perform my job in an efficient manner	16

Chapter XII

Safety and Health in the Virtual Office

Jay T. Rodstein and Katherine S. Watters
Honeywell Technology Center, USA

ABSTRACT

Safety and health issues in virtual offices are part of progressive telecommuting programs. Telecommuting agreements between employers and employees typically include clauses related to keeping the workspace free from hazards, conditions to clarify employer liability for injuries and illnesses and mechanisms for the employee to report them. In practice, employers are not providing all of the tools, training and follow-up needed to offer complete programs. This chapter provides a set of tools and implementation strategies to help employers reduce safety and health risks in the virtual office.

INTRODUCTION

Virtual office use is increasing for many reasons. Telework can benefit society by decreasing traffic congestion and vehicle emissions, thus reducing the number of traffic accidents and the amount of air pollution (Johnson, 1998). Employees may benefit through improved morale, reduced stress, and help in balancing work and personal life. Employers may benefit through increased productivity, reduced costs and improved recruitment and retention of valuable employees (Nilles, 1998). Statistics point to a rising trend of telecommuting as a work option in the United States. According to Cyber Dialogue/FINDsvp's American Internet Users' Survey (Telework America,

1999), the number of US teleworkers rose from 4 million in 1990 to 9.1 million in 1994 and to 15.7 million in 1998. Today's technology and communication services have made it easier for individuals to communicate and work from anywhere.

Along with the benefits of telecommuting, there are safety and health risks associated with virtual offices that must be understood and addressed in a complete telecommuting program. The initial emphasis in teleworking, as with other new work methods, is on its benefits to the employer, specifically production and efficiency gains. In many cases, safety and health are considered in the program development however, follow-through in training, workstations design, and inspection is less likely. Often, identification, evaluation, and control of safety and health risks are part of a second wave of process improvements, which incorporate initially unaccounted costs. Human costs, such as chronic or delayed effects associated with repetitive stress injuries (RSIs), are not included in the initial evaluations of telework.

This chapter concentrates on the at-home office, but many of the observations and recommendations apply to other virtual office environments. The first objective is to identify safety and health issues in the virtual office environment. The risk of RSIs due to poor ergonomic tools, physical arrangements and work practices of the workstation in the at-home office are emphasized, but other physical hazards in the virtual office also are discussed. Next, an approach to reducing the hazards to employees and costs to employers through the development of a set of intellectual tools, physical tools, and an implementation strategy is presented. Finally, suggestions for safety and health improvements for the increasing number of virtual office workers are provided. Concerns center on the following three issues:

1. Inclusion of less-than-ideal employee candidates, as the number of telecommuters increases.
2. Inadequate follow-ups by employees to ensure tools are being properly used.
3. Inadequate ergonomic features for at-home office furniture.

BACKGROUND

Facts about Repetitive Stress Injuries and the Office Worker[1]

OSHA states that *work-related musculoskeletal disorders* (WMSDs) are now a leading cause of lost-workday injuries and workers' compensation costs. Carpal tunnel syndrome, one form of WMSDs, leads on average to more days away from work than any other workplace injury (Occupational Safety and Health Administration, 1997). According to UNUM Corporation, a major disability insurer, carpal tunnel syndrome claims increased 467% from 1989 to 1994 (Leavitt, 1995). The National Council on Compensation Insurance contends that the lost wages and medical expenses for an average *cumulative trauma disorder* (CTD) claim are $29,000. A minor CTD case can cost $5,000-$10,000. Major cases requiring surgery may cost $100,000 or more with all indemnity considered (Leavitt, 1995). The cost of ergonomic-related injuries can be considerable, as is the cost saving that is calculated by having a proactive approach to such issues.

Figure 1 compares the workstation evaluations and furniture installation with OSHA claims for office-work-related *repetitive stress injuries* (RSIs) for Honeywell Technology Center (HTC). As workstations were evaluated and furniture was adjusted properly, ergonomic claims greatly declined. In 1998, reports of office work-related RSIs increased for HTC. This increase may reflect employees' greater awareness as a result of center-wide ergonomic training in 1997. This training included the causes and symptoms of RSIs, self-evaluation of workstations and stressed early reporting of symptoms.

Statistics on employees who had symptoms, but not necessarily worker's compensation claims, were kept for 1996-1999 (through August 1999). Follow-ups were done with the employees who had workstation evaluations and/or new adjustable furniture installed. Of all the symptomatic employees,

Figure 1: OSHA claims for repetitive stress injuries

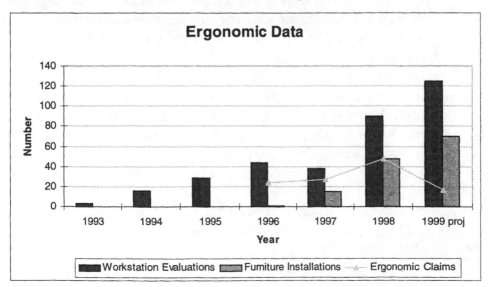

38% responded that the symptoms had completely resolved, 54% responded that the symptoms were mostly resolved and 5% responded that the symptoms had not resolved. Clearly these results show that symptoms related to the furniture setup can be curbed and almost eliminated, especially if caught early, by proper workstation adjustment and employee awareness. "It is therefore reasonable to assume that the decrease in physical discomfort reported at the adjustable workstation was due to both the preferred settings and the proper chairs." (Kroemer & Grandjean, 1997).

Proactive ergonomic programs provide maximum financial return-on-investment:

- An analysis by the National Institute of Occupational Safety and Health (NIOSH) found the average performance of data entry personnel increased 24% at ergonomically designed workstations compared to traditional settings.
- Among computer operators at an insurance company, a 10% to 15% increase in transactions per hour was realized as a result of seating and workstation improvements. Ergonomic seating alone contributed to a 5% increase in productivity.
- In 1990, one firm (a healthcare insurer) spent an average of $20,000 each for 30 CTD claims. After investing in office ergonomic improvements, they received a $1 million dividend from their insurer as a result of greatly reduced workers compensations claims.
- A major California newspaper outfitted more than 500 computerized workstations with new ergonomic furniture and accessories, such as keyboard supports, document holders, footrests, and glare screens on monitors. In just the first year, medical claims for wrist disorders were reduced by 63% and employees moral increased significantly. (Leavitt, 1995, pp. 2, 4).

So what do all these organizations, numbers, and acronyms mean? Regardless of whether one believes RSIs are a scourge to office workers, the following can be stated: RSIs occur in the office environment, result in a high number of lost workdays, are more expensive than other occupational illnesses, and can be reduced with a complete program of education, workstation design, and follow-up.

Although there are few reported cases of RSIs in the at-home office, the same factors that cause RSIs to occur in the traditional office environment are present in the virtual office.

Responsibility for the At-Home Office

In the United States, the Occupational Safety and Health Act of 1970 requires employers to furnish employees with a workplace free from recognized hazards (OSHA, 1998). Some recent developments have sent mixed messages about OSHA's involvement and employer responsibility for at-home offices. On November 15, 1999, OSHA (1999a) provided guidance to an employer about their employees working at home. On January 5, 2000 (OSHA, 2000), after the compliance guidance letter came to the attention of the media, Secretary of Labor Alexis Hermann stated that OSHA would withdraw it. She said the letter went beyond the agency's mandate and that a dialogue between the government and employers must take place to adapt to the changing nature of work in the 21st century.

In testimony to Congress on January 25, 2000 (OSHA, 2000), OSHA Administrator William Jeffress stated that OSHA does not believe the employer is liable for the home work environment, although he reinforced that the employer is responsible for all work-related injuries and illnesses wherever they take place.

What to make of these recent comments regarding employer liability and prudent practice? Two key phrases from these current statements by OSHA are: "An employer is responsible for providing a safe and healthful workplace, not a safe and healthful home" (OSHA, 1999b) and "(Employers subject to OSHA Injury and Illness reporting requirements) continue to be responsible for keeping such records, regardless of whether the injuries occur in the factory, on the road, in a home office or elsewhere, as long as they are work-related" (Jeffress, 2000).

As with other occupational safety and health standards, states may be more stringent than the federal requirements. The State of California adopted an ergonomics standard in July 1997 (California Code of Regulations, 1997). This law has been contentious, and parts of the standard have been challenged in state court, but the bulk of the standard is in full effect for most employers. The California program requires employers who have had more than one injury caused by an identical repetitive motion activity to have a program requiring analysis of the offending job, control of the repetitive task(s), and training. Other states, including Arizona, Minnesota, and Oregon, have telecommuting programs, including safety and health provisions, for state agencies and guidance for potential telecommuting companies on the Internet.

A majority of states provide for employer liability under workers' compensation wherever the injury occurs, provided the injury arose out of, or is related to, employment and the injury occurred in the course of employment (Head, 1997). For this reason, it is necessary to specify conditions of employment, in as much detail as is feasible, in a telecommuting agreement.

Local codes and ordinances may impact the environmental, health and safety requirements for an at-home office. These may include zoning issues or fire codes; for example, are fire extinguishers, smoke detectors or emergency plans required? The fire marshal can be contacted regarding the home/work environment and specific requirements for the at-home office. In many cases, the fire official will be in a jurisdiction different from the company offices and zoning and fire codes may not be the same. Insurance considerations must be taken into account. Employers should make sure they are covered for business loss and liability associated with at-home office duties. Employees should ensure that their insurance policy includes the use of an at-home office and that coverage is sufficient for these needs. A telecommuting agreement should include a provision addressing employer and employee liability.

This book looks primarily at issues in the United States. However, multinationals, non-US companies, and companies employing persons outside the United States should consider host country requirements. For example, in the United Kingdom, the Health and Safety (Display Screen Equipment) Regulations of 1992 place certain requirements on employers, including providing equipment to a certain minimum standard, assessing workstations, and providing information, instruction and training (Budworth, 1996). In New South Wales, Australia, a first aid kit is required in all workplaces, including home-based businesses, to meet the State's Occupational Health and Safety Regulation, according to the Office of the Director of Equal Opportunity in Public Employment (1995).

Most successful telecommuting organizations have accepted the premise that teleworkers' offices are covered by the same rules as the principal office (Nilles, 1998).

Workstation Design

Attention to ergonomics is especially important in a telecommuting program. "The need to provide a safe, healthy and productive environment in the home is paramount. The application of sound ergonomic principles is an integral part of meeting these needs. Ergonomics is the science that explores the relationships between people, their equipment and the tasks they perform. Ergonomics looks at these factors in relation to the productivity and comfort they provide, all within the framework of human physical and psychosocial issues. When ergonomics is neglected there is a potential for a mismatch between job demands and what the human mind and body can provide; this mismatch threatens the health and safety and productivity of the worker" (Hancock, 1999). Ergonomics has become a buzzword. It involves more than just purchasing furniture with purported *ergonomic features*. It is not just fitting the furnishing to the human being, it is fitting the workstation to the individual and the tasks performed.

Figure 2 compares two workstations: one which incorporates ergonomics into the design and one that does not. The chair is the foundation of the workstation. It is imperative that the chair be height adjustable. Preferably the seat back is adjustable and independent of the seat pan, providing good lumbar support. The chair must be adjusted to the proper height because all other adjustments will depend on the height of the chair. To find the correct height adjustment, the telecommuter should sit at a height where the feet are flat on the floor, with no pressure on the back of the thighs. The chair's back support should be adjusted to the lumbar area of the employee's back. Whether there are armrests on the chair is personal preference. They do support the forearm, but are not necessary. If there are armrests, they should be adjusted to support the arm when it is resting without applying pressure to the elbow or forearm when the telecommuter is keying or using a pointing device.

It is especially important that the keyboard be at the correct height. It is likely that a standard desk surface will be too high. The forearms should be parallel to the floor and the wrists maintained in a neutral position. If the keyboard surface is too high or too low, the wrists will be forced into a flexed or extended position. To determine the correct height for the keyboard

Figure 2

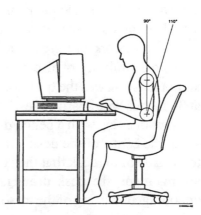

surface, whether this is an adjustable keyboard tray or the actual surface of the desk, the employee should sit in the chair after it has been adjusted to the correct height. The employee should relax his/her elbows at his/her sides and extend the forearms until they are parallel with the floor. The distance from the elbow to the floor should be measured. This is the height that the home row of the keyboard should be set (Nilles, 1998). The pointing device should be at the same height as the keyboard and close enough so the employee does not have to extend the shoulder or forearm to reach the mouse hand. The best location for the mouse is adjacent to the keyboard on the same surface.

The top of the monitor should be at eye height or, if the monitor is large, the center of the monitor should be adjusted to the center of the vision. Those who wear bifocals should position the monitor lower so they can maintain a natural neck posture when looking down through their lenses. "This arrangement accommodates the resting condition of the eye and avoids visual fatigue. For this reason one should be able to position a display so that the viewing angle is about 45 degrees below the eye-ear line" (Kroemer & Grandjean, 1997). The monitor should be placed approximately an arm's length from the face.

The desired workstation layout is the cockpit configuration. In this configuration the computer is placed in the corner and the work surface extends to either side. Compared to the straight desk and L-shape, this allows the worker to access a greater area of the work surface without reaching (Anderson & Hoseck, 1999). It is important that there are no barriers to movement, such as table legs.

The following is an overview of a work-related injury, which demonstrates the importance of workstation design. The claim was related to an incorrect workstation setup resulting in strain to the right hand, arm, shoulder, and hip. The employee believed cumulative trauma was sustained while cradling the phone between the neck and shoulder when answering phone calls and that an inadequate chair was provided. In 1996, six years after the initial claim, the case was settled for tens of thousands of dollars. As a result of that case, workstation evaluations were performed proactively for employees who complained of discomfort. At the time, the most common worksurface at that location was the traditional desk. The desk had a writing surface fixed at 30 inches with sharp edges. The file drawers were located on either side of a center opening with a pencil drawer. In the best circumstances, the monitor sat in the center of the desk. Often, the employees had a table perpendicular to the side of the desk that they used to write or organize their projects. In this arrangement, the desk drawers presented a barrier to leg movement and encouraged twisting of the torso. Another popular configuration was to angle the monitor and keyboard on one side of the desk. This arrangement freed the desk area directly in front of the seated employee for writing and reading. However, this setup resulted in employees twisting at the waist, neck and shoulders to view the monitor and compressing the ulnar and median nerves on the sharp edge of the desk. In a desperate attempt to get the keying and writing surfaces to the correct heights, the occupational health nurse had the legs cut down on the desks to lower the writing height, and removed the center pencil drawers to install adjustable keyboard trays. This temporary fix did not solve all the problems, but it got the management's attention.

At this point, with management support, an aggressive office ergonomics program was initiated that included training, assessment tools, medical management of symptomatic employees, and installation of adjustable furniture over a five-year period beginning in 1997. Preceding the furniture upgrades, all employees attended an organizationwide ergonomic education program to bring awareness to proper posture and working habits.

As the office ergonomics program was being implemented, a trend of informal telecommuting (mainly employees working out of their homes) developed. These informal arrangements were typically agreed upon between employees and supervisors. In some cases, company equipment was provided for the at-home office.

One of the informal arrangements resulted in an occupational illness. The employee was working at home, proofreading and correcting a document. The employee reported working at a kitchen table, leaning on the left elbow for 40 hours over five days. The diagnosis was ulnar nerve neuritis. The employee was treated for this condition for six months, at which time surgery was recommended and performed. Recovery from surgery kept the employee out of work for another month, at which point follow-up with a physician continued until maximum medical improvement was reached nine months later.

These types of safety and health issues, as well as factors such as cost savings from reduced floor space needs and a desirable work/life option, led to the development of a formal telecommuting program. After a year of planning, the formal program was introduced to employees in December 1998. The program includes a strong training component and employer/employee cooperation in selecting the at-home office location and furniture. The company evaluates the employee's own furniture for ergonomic adequacy or provides funding for basic workstation furniture and accessories through an approved vendor. Vendor installation of company-provided equipment is required. A formal telecommuting agreement is in place that includes provisions for maintaining a workplace free from safety hazards, clearly states the responsibilities associated with injuries and reporting, and gives the company the right to inspect the at-home office.

Benchmarking

Given this background, it is not surprising that the benchmarking Honeywell Technology Center has done with other companies that have active telecommuting programs shows most employers incorporate a safety and health component into their telecommuting programs and place a high emphasis on workstation design and ergonomics. Honeywell Technology Center benchmarked a number of safety and health criteria of telecommuting programs with six other companies. Key findings include the following:

1. Five of seven companies had formal telecommuting agreements that addressed the requirements for maintaining a safe work environment, the right for the employer to conduct a home inspection, and the responsibility for injuries that occur in the home.
2. Workstation design was acknowledged as a critical component; however, the methods used to incorporate this varied from employee-driven to employer-driven processes.
3. Furniture offerings were considered adequate and accepted by the employees.
4. Variable levels of financial support and methods of support were reported, ranging from no support to complete funding and installation of furniture.
5. The employer retained the right for home inspection in all formal programs, but in all but one case, inspections were not conducted. No Company had a routine, recurring inspection program in place.

ISSUES AND SOLUTIONS

General office safety and in particular prevention of RSIs are an issue for virtual office workers, who are performing the same tasks as traditional office workers with less direct contact with support and supervision. How should employers integrate safety and health into a telecommuting program? The answer requires providing the proper intellectual, physical tools and having a formal implementation strategy.

Intellectual Tools

The key player in a successful telecommuting program is the supervisor. Before the first teleworker keys a stroke in the virtual office, the supervisor must understand the differences in managing teleworkers. From a safety and health perspective, this means greater responsibility for the employees' welfare, given that the telecommuter will have less direct contact with company support functions and peer contact. Supervisors must regularly ask about safety and health issues. These may include simple queries about the status of furniture installation or repairs, reminders to take breaks, and direct questions about symptoms. The supervisor must receive the same or greater level of basic training in safety and health topics as teleworkers.

From a safety and health perspective, minimal initial training would include the ergonomics of workstation design and use, accident reporting, evacuation planning, and general office safety. In businesses where these topics are covered for traditional office employees, a refresher with specific examples for the at-home office environment is crucial. In particular, self-evaluation of the workstation and early reporting of symptoms in the at-home office should be stressed. Examples of questionnaires, which can be used for training, self-evaluation and inspection, can be found in Budworth (1999), Keller (1999), Roads and Traffic Authority (1998) and Joice (1996). RSIs have an insidious onset, and employees may ignore or rationalize the symptoms and not take the time to notify the health care contact or supervisor. Early intervention has been successful at keeping employees healthy and working. The following case study shows how ergonomic education, awareness of symptoms, and early reporting of problems produced a positive outcome.

An employee complained of right elbow pain, extending into the forearm and wrist. The symptoms had started "some time ago" but had worsened during the previous two weeks. The employee reported that the symptoms developed within 10 minutes of using the mouse. A workstation evaluation one year earlier had resulted in the installation of an adjustable keyboard tray. The employee participated in ergonomic training and reported that ergonomic changes also were made to his at-home workstation. A visit to his work environment revealed the cause of the pain in his elbow and hand. The fixed-height keyboard tray had loosened and lowered over time; consequently, the keying and pointing surface was too low. The elbow had constant pressure while mousing because it rested on the edge of the chair arm. Due to the employee participating in ergonomics training, reporting his symptoms early, and implementing modifications and treatments, his symptoms completely reversed without outside medical assistance. If this employee had continued to cause inflammation to the elbow until surgery was indicated, the cost of this claim could easily have reached $30,000.

More basic than understanding ergonomic principles and more frequently overlooked is training in the adjustment of furniture and chairs in particular. It may sound trivial, but the typical office chair frequently remains unadjusted despite its several adjustable features and multiple levers to make these adjustments. HTC health professionals have found that many employees have difficulty understanding the value of these features and coordinating the adjustments of the "common" office chair. In the traditional office, where assistance is accessible, employees only ask for help when they become desperate. Therefore, initial and refresher training should be provided to telecommuters, including hands-on training with model equipment in the traditional office and with the actual product installed in the at-home office.

Physical Tools

"Many factors come together to provide an effective home office. The most important ergonomics factor in the design process is recognizing the incredible diversity that exists among people; the normal person based on data tables simply doesn't exist. In this light, success of the home office in the ergonomics realm equates to the degree of adjustability and flexibility inherent in the workstation and equipment." (Anderson & Hoseck, 1999).

Benchmarking with other companies indicates that where there is a formal telecommuting program, at a minimum, chairs are supplied. Most programs provide work surfaces as well as some accessories, including surge suppressors and keyboard trays. Employers provide a wide range of support, from all-inclusive to leaving employees on their own in matters such as work surfaces, chairs, storage, communications, electronics and electrical requirements. Typically employers will pay for additional data connection points; however, all other physical changes to the at home office space, including additional electrical outlets, doors, noise dampening, and climate control, are expected to be the employee's responsibility.

Regardless of who is providing the furniture for the home office, it is important that the chair be ergonomically adequate, as the chair is the foundation of the workstation. A review of what is on the market revealed that the selection of chairs is good, featuring a lot of adjustability. However, workstations for the home that are adequately adjustable are hard to find, despite the considerable use of the term *ergonomic* in advertising. Several features that should be sought in a workstation furniture system are:
1. Ability to set keying and writing heights.
2. A pointing device to be positioned adjacent to and at same height as the keyboard.

3.   A rounded edge profile on work surfaces.
4.   Corner units to which a worksurface may be attached to either side (left-handed and right-handed). The corner unit also allows the user to easily reach items on the desk to either side.
5.   Minimum barriers (such as table legs) allowing movement throughout the work area, minimizing twisting.

Office furniture vendors were questioned about incorporating the above desired features into the furniture lines they offered for the home office, such as a worksurface with height adjustability and barrier free systems. The response was that their customers do not request such features and adding these features would increase the cost of the furniture. This suggests that people responsible for selecting furniture, particularly furniture for telecommuters, need to understand the tradeoffs between the cost of RSIs and the cost of the needed features.

Besides the ability to adjust the workstation, cooperation between employee and employer is needed in selecting a suitable home office location. Other important safety considerations include electrical safety, fire protection and general housekeeping to prevent slips, trips and falls. Employers must consider local fire codes as well as state, county, or city ordinances as they relate to safety and health issues. While many believe that virtually any private location in a home is suitable for telecommuting, the emphasis has been on the work and not necessarily on the worker. Many basements, attics, and other spaces may lack environmental controls for heat, humidity, ventilation, noise, and light. They may also be physically unsuitable, lacking clearance around furnishings and electrical or data transmission equipment to support the at-home office. Employees must receive guidance on optimizing the location and layout of the at-home office. Example checklists that can be used to select an appropriate home office location can be found in Budworth (1999), Keller (1999), Roads and Traffic Authority (1998) and Joice (1996, Appendix C).

Implementation

Once the level of employer support has been determined, training has been developed, and the furnishing program details have been determined, a process should be implemented to ensure both telecommuter's and employer's needs are met. The first step is for prospective telecommuters and their supervisors to agree that a telecommuting arrangement is desirable. This assessment should include direct contact with a representative of employer support functions. For example, prospective telecommuters may meet with representatives from the Information Technology, Human Resources, Facilities, Property, and Safety & Health departments. This may be one person in a small firm or many people in a larger company. These contacts ensure that teleworkers understand the issues for at-home work, are provided with the proper tools to make at-home work successful, and have a person to contact in addition to the supervisor regarding specific functional issues. In our experience, it is insufficient for the telecommuter's supervisor to serve as the sole point of contact. The supervisor may not be objective about the telecommuter's needs, may not be aware of the full range of solutions, and may make undesirable short-term cost saving tradeoffs. However, the supervisor has the most direct contact with the telecommuter and is the key player in assessing operational success, including the safety and health provisions of the program.

The preliminary assessment concludes with the supervisor and telecommuting candidate signing a telecommuting agreement. This agreement should include provisions to maintain a hazard-free workplace and to provide resources to ensure this can be accomplished. An inspection clause, which allows the company to ensure a safe and secure workplace, should be included as well as procedures for injury and illness reporting. This may be as simple as stating that procedures are the same as in the traditional office; however, spelling out work hours, home office access restrictions, and business visits are included in many agreements. Examples of safety and health provisions in telecommuting agreements can be found in Gordon (2000) and Office of the Director of Equal Opportunity in Public Employment (1995).

A critical component to ensure the telecommuting arrangement gets off to the right start is the at-home office evaluation. The employer should conduct an at-home office inspection if possible to ensure the conditions in the agreement can and are being met. If an at-home inspection is not performed, the telecommuting employee should fill out a self-evaluation checklist. Several references for these checklists were provided earlier in the chapter. Once these steps have been completed the process does not end. Continuing contact by the supervisor and possibly other specialists to assess safety and health must be built into the program to ensure that symptoms do not go unnoticed and that safety practices are equivalent to the traditional office.

FUTURE TRENDS

Currently, employers are selective about the employees who become telecommuters. This may be causing an optimistic picture of the injury and illness, and workers' compensation experience. As telecommuting becomes more mainstream, expected by workers as an option and encouraged by employers as a cost-saving and efficiency tool, the expanded labor pool of telecommuters may not exhibit the high ideals that the more select pool presently displays. This increased number of telecommuters will require a standard approach with less attention to individual telecommuters and their specific work environments and greater understanding by supervisors and telecommuters of at-home safety and health issues.

Progressive companies with telecommuting programs are doing a very good job of providing the needed tools to employees, but only a fair job of ensuring that those tools are being used correctly. The tools are only as good as the employee's ability to use them. This requires effort on the part of telecommuting trainers to ensure that employees and supervisors understand the benefits and costs of proper installation, maintenance, and use of furnishings. Supervisors are responsible to assess the safety and health conditions of their telecommuting employees even more than employees at the traditional office. Telecommuters are responsible for using the safety and health tools for their own benefit and for staying productive. In HTC's experience and that of the benchmark partners, inspections to ensure that the physical environment and work process meet safety and health criteria are not done routinely. The telecommuting program is only as strong as its weakest component; therefore, unless inspections are conducted at the time the office is put to use and follow-up inspections are done routinely, the safety and health benefits may not be realized.

Furniture offerings for the at-home office are inadequate. They lack the set of ergonomic features needed to provide equivalently safe, at-home workstations. Specific features that are missing are adjustable work surfaces, consistent rounded edges, and barrier-free furniture systems. Currently, vendors are providing a price-competitive product based on incomplete costs. Once telecommuting companies recalculate costs to include human and productivity losses due to illness, they will demand these features from furniture vendors. Once vendors are driven by the market to produce these improved offerings, prices may still be higher than current offerings, but efficiencies in production and competition for sales will bring them down.

CONCLUSION

Concern for worker safety and health is an integral component of the modern workplace. The benefits of telecommuting are great and it is gaining acceptance as a work option. The telecommuter's office is subject to the same safety and health criteria as the traditional workplace and these issues must be incorporated into progressive telecommuting programs. Telecommuting agreements typically include clauses related to keeping the workspace free from hazards; setting forth conditions to clarify employer liability for injuries and illnesses as well as mechanisms for the employee to report them. A complete telecommuting program must include risk management decisions regarding safety and health as well as other business factors. Progressive employers realize that a complete program includes providing the proper tools, training in how to use the tools, and ensuring the tools are used properly.

REFERENCES

Anderson, M.A., & Hoseck, L. (1999). *Hennepin County Televillage – Home Office Ergonomic Guidelines.* Minneapolis, MN.

Budworth, N. (1996). Teleworking—out of 'site', out of mind? *The Institution of Occupational Safety and Health Datasheet* [On-line]. Available: http://www.iosh.co.uk/inform/datasheet/teleworking.doc. [April, 1996].

Budworth, N. (1999). *Assessment of premises for teleworking. The Institution of Occupational Safety and Health Datasheet* (On-line). Available: http://www.iosh.co.uk/inform/datasheet/teleworking.doc. (February 26, 1999).

California Code of Regulations. (1997). Title 8-Sect. 5110, *Repetitive Motion Injuries (RMIs)* [On-line]. Available: http://www.dir.ca.gov/title8v/5110.html. [July 3, 1997].

Gordon, G. (2000). Generic Telecommuter's Agreement [On-Line]. Available: http://www.gilgordon.com. [February 28, 2000].

Hancock, P.A. (1999). *Human performance and ergonomics.* Boston: Academic Press.

Head, C. A. (1997). *Telecommuting: Panacea or Pandora's Box?* Law Watch Publications, Holland and Knight LLP [On-line]. Available: http://www.hklaw.com/publications/attorney/heada3111997.html.

J.J. Keller & Associates (1999, March). Checklist: office area. *Keller's Industrial Safety Report*, p. 16.

Jeffress, C. N. (2000). Testimony before the Subcommittee on Employment, Safety, and Training, Senate Committee on Health, Education, Labor, and Pensions. [January 25, 2000].

Johnson, R. P. (1998). *Safety and health benefits of telecommuting.* Telecommuting Safety and Health Benefits Institute [On-line]. Available: http://www.orednet.org/venice/rick/telecommutesafe/safework2.html. [April 1, 1998].

Joice, W. H. (1996). Interagency telecommuting program implementation manual, appendix C: Self-certification safety checklist for home-based telecommuters. U.S. Government Services Administration, January 1996.

Kroemer, K. H. E. & Grandjean, E. (1997). *Fitting the task to the human – a textbook of occupation ergonomics.* Philadelphia, PA: Taylor & Francis Ltd.

Leavitt, S. B. (1995). Return-on-Investment. *Ergonomics in the Office, 6.* Mead–Hatcher, Inc.

Nilles, J. M. (1998). Managing Telework: Strategies for managing the virtual workforce. New York: John Wiley & Sons.

Occupational Safety and Health Administration (1997). News Release USDL 97-146. OSHA Launches Ergonomics Web Page on Workers' Memorial Day.

Occupational Safety and Health Administration (1998). *Occupational Safety and Health Act, Introduction.* (Public Law 91 – 596, Sect. 2193, December 29, 1970, as amended by Public Law 101-552, Sect. 3101, November 5, 1990, as amended by Public Law 105-198, July 16, 1998, as amended by Public Law 105-241, September 29, 1998). Washington, DC: U.S. Government Printing Office.

Occupational Safety and Health Administration (1999a). OSHA Standards Interpretation and Compliance Letter. OSHA policies concerning employees working at home. November 15, 1999.

Occupational Safety and Health Administration (1999b). *Working Draft of OSHA's Proposed Ergonomics Program Standard* [On-line]. Available: http://www.osha-slc.gov/SLTC/ergonomics/backgroundinfo.html. [November 20, 1999].

Occupational Safety and Health Administration (2000). OSHA Press release (DOL:00-05), Statement of Secretary Alexis M. Herman, U.S. Department of Labor. January 5, 2000.

Office of the Director of Equal Opportunity in Public Employment (1995). *Success with flexible work practices* [On-line]. New South Wales, Australia. Available: http://www.eeo.nsw.gov.au/frampubl.htm. [October 1995].

Roads and Traffic Authority (1998). *How to set up a teleworking program (Issue 2.0)* [On-line]. New South Wales, Australia. Available: http://www.rta.nsw.gov.au/frames/initiatives/e_f.htm?/frames/initiatives/e4&/initiatives/e44_c.htm&How+to+set+up+Teleworking&4. [March 23, 2000].

Telework America (1999). *Telework America Overview.* [On-line]. Available: http://www.telecommute.org. [March 23, 2000].

ENDNOTE

1 In this section, several terms are used to describe illnesses associated with repetitive motion. The Occupational Safety and Health Administration (OSHA) uses the term *illness* as distinguished from *injury,* because the effects occur over time, whereas injuries result from trauma. The terms *repetitive stress injuries* (RSIs), *work-related musculoskeletal disorders* (WMSDs), and *cumulative trauma disorders* (CTDs) describe a group of similar illnesses that include, but are not limited to, repetitive typing and keying illnesses. In this chapter they are referred to as RSIs.

Chapter XIII

Telecommuting Experiences and Outcomes: Myths and Realities

Donna Weaver McCloskey
Widener University, USA

ABSTRACT

There are many contradictions concerning expected telecommuting experiences and outcomes. At one extreme telecommuting is believed to benefit the employee by providing increased flexibility and job satisfaction and reduced stress. On the other extreme telecommuting has been said to result in very negative experiences and outcomes for employees including isolation, increased stress and limited career advancement opportunities. This research attempts to separate the telecommuting myths and realities by examining the impact of this work arrangement on work experiences and outcomes for professional employees. This research found telecommuting experiences and outcomes are largely positive for professional employees. The telecommuters reported significantly more autonomy, boundary spanning activities and career advancement prospects and significantly less time and strain-based work-family conflict than their non-telecommuting peers. The only negative experience that was found was that telecommuters received less career support than non-telecommuters. The lack of career support did not hinder career advancement prospects.

INTRODUCTION

Rarely has a work arrangement inspired as much anticipation and trepidation as telecommuting, a work option in which organizational employees work from home in lieu of the traditional workplace. Advertisers sell the flexibility and life balance that telecommuting will bring while editorialists forecast the isolation, overwork and stress that comes from the ability to work anytime, anywhere. Even researchers are divided on the potential outcomes of telecommuting. It has been suggested that telecommuting will increase flexibility and job satisfaction and reduce employee stress (DuBrin & Barnard, 1993). Conversely, other authors have implied that telecommuting has negative outcomes for employees, including loneliness and isolation (Lewis, 1997), increased stress and limited career advancement prospects (DuBrin & Barnard, 1993). Despite the mixed opinions concerning the outcomes of telecommuting, this work arrangement has experienced a tremendous amount of growth. According to a survey conducted by CyberDialogue, there were 15.7 million telecommuters in the US in 1998 (Gordon, 1998). It has been estimated that as many as 51% of all North American companies allow employees to telecommute through ongoing or pilot programs (McShulskis, 1998), and expansion is expected to continue. With the continued growth of this work option, it is imperative that the impact of telecommuting on work experiences and outcomes be addressed so that organizations can maximize the potential benefits and minimize the potential pitfalls. This research attempts to separate telecommuting myths from reality. This process begins by briefly examining why there are so many contradictions concerning telecommuting.

The primary cause for the numerous contradictions in telecommuting research is methodological issues. (For an in-depth discussion and an expanded literature review, please see McCloskey & Igbaria, 1998.) First, the definition of telecommuting varies widely. Farmers, self-employed people who work at home, people who manufacture shoes from home, business people who work at home in the evening and organizational employees who work at home in lieu of the office have all been characterized as "telecommuters." Work experiences and outcomes would be expected to differ for such a diverse group, yet all too often they are combined under the single "telecommuting" banner. The second reason for contradictions in telecommuting research is that many studies fail to control for significant extraneous variables. Researchers have analyzed telecommuters as a homogeneous group, regardless of whether the employees are full-time or part-time; whether they are blue-collar, clerical, managerial or professional employees; or whether telecommuting is done on a full-time, part-time or occasional basis. These employees should not be analyzed as a homogeneous group because their differing characteristics may contribute to differences in their experiences and outcomes. This research attempts to address these concerns by using a focused definition of telecommuting and limiting the sample to professional employees who regularly work at home instead of the office one or more days per week. This definition focuses on employees who telecommute to substitute for working in the office as opposed to those who work at home in the evenings to supplement their full-time office work.

This research explores whether there are differences in the work experiences and outcomes of professionals who choose to telecommute and those who choose to work in the traditional office environment. By using a focused definition of telecommuting and a sample limited to professional employees, this research attempts to separate the myths and realities of telecommuting experiences, including autonomy, work-family conflict, career support and boundary-spanning activities and work outcomes, including organizational commitment, job satisfaction and career advancement opportunities.

Work Experiences

Autonomy is the degree to which a job provides freedom, independence and discretion in the completion of work. Telecommuting is generally believed to increase autonomy by providing flexibility and control over the completion of work. Professional employees in the traditional office environment may have control over the prioritization of work. If these employees choose to telecommute, they will experience additional autonomy. The employees still have control over the prioritization of work, but now the employees may also choose when and where the work will be completed. Others have argued that telecommuting actually results in less autonomy because managers will be more controlling of employees who work outside the traditional work environment (Olson & Primps, 1984). Although it generally is believed that telecommuting provides professional employees with additional control and discretion over the completion of work, this has not been empirically established.

Hypothesis 1: Telecommuters will experience more autonomy than non-telecommuters.

Work-family conflict is a form of conflict in which the role pressures from work and family domains are mutually incompatible. The perceived impact of telecommuting on work-family conflict is quite diverse. Many believe that telecommuting will provide the flexibility to meet demands from both work and family (DiMartino & Wirth, 1990). Others have suggested that telecommuting will make it difficult to separate work and family, thus increasing conflict and stress (DuBrin & Barnard, 1993). These views are not necessarily contradictory. Researchers have proposed that work-family conflict is comprised of time, strain and behavior-based conflict (Greenhaus & Beutell, 1985). It is possible that telecommuting will impact these three dimensions differently, thus decreasing some dimensions of work-family conflict and increasing others.

Time-based conflict suggests that the demands of one role make it difficult to be either physically or mentally available to the other role. Professionals may experience this form of work-family conflict when their travel commitments make them miss an important family event. It is this dimension of conflict that telecommuting helps relieve. By giving employees more control over when work is completed, the employees should be able to schedule work to minimize the conflict between work and family demands.

Hypothesis 2: Telecommuters will experience less time-based work-family conflict than non-telecommuters.

Strain and behavior-based work-family conflict involve the ability to make adjustments between the work and family role. Strain-based work-family conflict is when the strain from one role makes it difficult to comply with the demands of another role. For example, strain-based work family conflict may occur when employees have an awful day at work and therefore are miserable when they come home in the evening. Behavior-based work-family conflict is when the behavior expectations of one role are incompatible with the other role. For example, employees may need to be tough and demanding at work, yet their families expect them to be warm and nurturing. Both of these types of conflict involve making a transition from one role to another. Commuting time may provide employees with transition time (Young & Morris, 1981) by giving the employees time to change their attitude and behavior to the role they are entering. Because telecommuters have eliminated their commutes, they are forced to change roles much faster, which may result in increased strain and behavior-based work-family conflict.

Hypothesis 3: Telecommuters will experience more strain-based work-family conflict than non-telecommuters.

Hypothesis 4: Telecommuters will experience more behavior-based work-family conflict than non-telecommuters.

It has been suggested that employees who choose to telecommute will no longer be considered serious by their supervisors and will therefore receive less interesting and visible assignments, less feedback and little or no mentoring (Fitzgerald, 1994). Empirical research has not addressed whether professional employees who telecommute receive less career support than those who choose not to telecommute. Given the potential negative impact of limited career support on advancement opportunities, it is important to establish whether telecommuters and non-telecommuters actually receive different levels of career support.

Hypothesis 5: Telecommuters will have less career support than non-telecommuters.

Boundary spanning refers to individuals crossing intradepartmental and inter-organizational boundaries to perform their jobs. These activities may suggest a level of visibility both within the organization and to external constituents. It has been suggested that telecommuters may become isolated and have less visibility (Stanko & Matchette, 1994), which may imply they participate in fewer boundary-spanning activities. Limited boundary-spanning activities have been found to have a negative impact on organizational commitment (Baroudi, 1985) and job satisfaction (Keller & Holland, 1975). Given these undesirable outcomes, it is important to understand whether professionals who telecommute do experience fewer boundary-spanning activities then their non-telecommuting peers.

Hypothesis 6: Telecommuters will have fewer boundary-spanning activities than non-telecommuters.

Work Outcomes

Telecommuting may result in less organizational commitment. If telecommuters become isolated and no longer have social contacts at work, they may have less commitment to the organization (Connelly, 1995). Others have suggested that organizational commitment may increase in appreciation for being allowed to telecommute (Fuss, 1995). It is important to understand whether telecommuting has an impact on organizational commitment because this work outcome has a direct effect on turnover (Igbaria & Greenhaus, 1992). The following exploratory hypothesis is proposed:

Hypothesis 7: Telecommuters will have less organizational commitment than non-telecommuters.

Despite all of the potential negative experiences and outcomes related to telecommuting, it has been suggested that telecommuting will result in increased job satisfaction (Wright, 1993). Telecommuting employees may be satisfied with their jobs because their organization is allowing them to have the flexibility necessary to combine work with another valued goal. Researchers have found that managers believe telecommuting will result in increased job satisfaction (Bailey & Foley, 1990) but this has not been established empirically.

Hypothesis 8: Telecommuters will have more job satisfaction than non-telecommuters.

The "out of sight, out of mind" adage is frequently used to argue that telecommuters will have fewer career advancement opportunities than non-telecommuters. In addition to less visibility, telecommuters may no longer receive information about advancement opportunities and may, as proposed in Hypothesis 5, experience less career support than non-telecommuters. Others have argued that telecommuting may actually result is increased career advancement opportunities due to increased productivity from working out of the office one or more days per week (Riley & McCloskey, 1997). Although there has been some limited support for telecommuting having no impact on career advancement (Olson, 1989), the majority of the studies (e.g., Bailey & Foley, 1990; Stanko & Matchette, 1994) indicate that there is at the least the potential for telecommuting to negatively affect career advancement. The following exploratory hypothesis is offered to test this belief.

Hypothesis 9: Telecommuters will have less career advancement prospects than their non-telecommuting peers.

METHOD

Procedure and Sample

Telecommuting policies and experiences vary widely among organizations. For this reason, it was deemed reasonable to conduct this research with a sample from one large organization. A large, highly competitive telecommunications firm agreed to participate in this research. At the time the organization was introduced to this project, they were in the process of collecting data from employees in their California offices to see how they

were meeting Clean Air initiatives. A statewide survey found 225 professional employees were telecommuting. The organization provided the names
and internal mailcodes for these employees. All 225 telecommuting employees received a copy of the survey instrument. Useable questionnaires were
received from 83 telecommuters (37% response rate). Each telecommuter
was asked to identify a person in a similar position who does not telecommute.
A total of 144 non-telecommuters were identified. Useable responses were
received from 71 non-telecommuters (49% response rate).

It was first necessary to determine whether there were significant
differences in the demographic characteristics of the telecommuters and non-
telecommuters. As indicated in Table 1, the process of data collection resulted
in a sample of telecommuters and non-telecommuters from similar job types.
The chi-square analysis revealed that there was not a significant difference in
the distribution of job titles among telecommuters and non-telecommuters.
As indicated in Table 2, there were significant differences between
telecommuters and non-telecommuters in organizational tenure, gender and
marital status. On average, the non-telecommuters had more organizational
tenure. The non-telecommuter group was 68% male whereas the telecommuters

*Table 1: Comparison of job titles for telecommuters and non-
telecommuters*

	Telecommuters (N=89)	Non-Telecommuters (N=71)
Account Executive	1 (1%)	2 (3%)
Administrator	26 (29%)	15 (21%)
Analyst	1 (1%)	1 (1%)
Coordinator	4 (4%)	3 (4%)
Director	1 (1%)	1 (1%)
Engineer	6 (7%)	3 (4%)
Facility Surveyor	0 (0%)	2 (3%)
Manager	21 (24%)	17 (24%)
Network Design	0 (0%)	1 (1%)
Sr. Administrator	12 (13%)	5 (7%)
Sr. Designer	0 (0%)	1 (1%)
Supervisor	9 (10%)	4 (6%)
System Analyst	0 (0%)	3 (4%)
Technician/Technical Support	2 (2%)	5 (7%)
Unreported	6 (7%)	8 (11%)

Table 2: Demographic characteristics of telecommuters and non-telecommuters

	Telecommuters (N=89)	Non-telecommuter (N=71)	t
Age	46.09	46.59	-.461
Org. tenure (years)	20.84	22.72	-1.659*
Hour per week:			
Paid employment	46.97	46.71	.186
Commuting	5.19	5.21	-.038
Childcare	4.06	5.39	-.746
Eldercare	1.73	.94	1.053
Household chores	11.56	10.94	.475
Recreation	15.37	13.37	.979
Job tenure (years)	6.40	6.71	-.316
			Chi Square
Gender: Male	44 (49%)	48 (68%)	5.334**
Female	45 (51%)	23 (32%)	
Eldercare: Yes	23 (26%)	18 (25%)	.005
No	66 (74%)	53 (75%)	
Childcare: Yes	32 (36%)	34 (48%)	2.149
No	56 (64%)	37 (52%)	
Education			
high school	7 (8%)	10 (15%)	2.587
some college	45 (51%)	22 (49%)	
bachelors degree	22 (25%)	18 (26%)	
graduate degree	14 (16%)	7 (10%)	
Marital Status			
Married	58 (65%)	57 (80%)	6.069**
Unmarried, living			
w/ partner	9 (10%)	7 (10%)	
Unmarried, not living			
w/ partner	22 (25%)	7 (10%)	
Salary			
$30,001-45,000	7 (8%)	8 (11%)	1.332
$45,001-60,000	43 (48%)	36 (51%)	
$60,001-75,000	29 (33%)	20 (28%)	
$75,000-90,000	6 (7%)	3 (4%)	
$90,001-115,000	3 (3%)	3 (4%)	
$115,001-130,000	1 (1%)	1 (2%)	

$p \leq .05$* $p \leq .01$** $p \leq .001$***

were evenly split between males and females. More of the non-telecommuters were married. There were no significant differences between telecommuters and non-telecommuters in age, hours per week spent meeting work and personal responsibilities, eldercare and childcare responsibilities, job tenure, education and salary.

Measures

Autonomy. Autonomy was assessed with the two-item measure from the Job Design Survey, developed by Hackman and Lawler (1971) and four other items from Parasuraman, Greenhaus and Granrose (1992) (alpha = .878).

Work-Family Conflict. The measures used to examine these three dimensions were developed by Gaitley (1996). Time-based conflict was assessed with 12 items (alpha = .842), strain-based conflict with nine items (alpha = .821) and behavior-based conflict with six items (alpha = .799).

Career Support. Career support is the extent to which individuals perceive their immediate supervisor to be interested in and supportive of their career. The four-item measure, developed by Greenhaus, Parasuraman and Wormley (1990), asked respondents to indicate how often they experienced types of career support (alpha = .706).

Boundary-spanning Activities. Boundary-spanning activities was measured with four items from the measure developed by Miles and Perreault (1976) (alpha = .854).

Organizational Commitment. Organizational commitment was assessed with an abbreviated version of Porter, Crampon and Smith's (1976) Organizational Commitment Questionnaire (OCQ) (alpha = .732).

Job Satisfaction. Job satisfaction was operationalized by a three-item scale developed by Hackman and Oldham (1975) (alpha = .709).

Career Advancement Prospects. Career advancement prospects were assessed with a single item. Respondents were asked to indicate the likelihood of being promoted within the current organization on a three-item scale from one (slight chance for promotion) to three (very good chance for promotion).

Data Analyses

The primary aim of this study was to determine differences between telecommuters and non-telecommuters on work experiences and outcomes. As indicated earlier, there are demographic differences between the telecommuters and non-telecommuters on organizational tenure, gender and marital status. It was therefore necessary to control for these factors so that conclusions regarding differences between telecommuters and non-telecommuters were not confounded by the demographic differences between these groups. Multivariate Analysis of Covariance (MANCOVA) was used to test for differences between telecommuters and non-telecommuters in work experiences and outcomes. MANCOVA allows for testing differences in the means of groups after controlling for initial differences between the groups. In this case, MANCOVA was used to determine whether there are differences in the work experiences and outcomes between telecommuters and non-telecommuters after controlling for differences in demographic characteristics. In each of the MANCOVA analyses, work arrangement was the independent variable and the work outcomes and experiences were the dependent variables. Organizational tenure, gender and marital status were the covariates.

RESULTS

Table 3 shows the results of the MANCOVA examining differences in work experiences between telecommuters and non-telecommuters. Three of the six proposed hypotheses concerning work experiences were supported. After controlling for demographic variables, the analyses indicate that telecommuters do experience more autonomy and less time-based work-family conflict and career support than non-telecommuters. Hypotheses 3 and 4, which proposed that telecommuters would experience more strain-based and behavior-based work-family conflict than non-telecommuters, were not supported. Contrary to the proposed hypothesis, non-telecommuters actually

Table 3: Results of MANCOVA for work experiences

Variables	Telecommuters (N=89)	Non-telecommuter (N=71)	F
Autonomy	4.0567	3.9779	3.695*
Work-Family Conflict:			
Time-based	2.1396	2.1717	4.808**
Strain-based	1.9557	1.9990	3.789*
Behavior-based	2.3236	2.1155	0.893
Career Support	2.8894	2.9029	5.178**
Boundary-spanning Activities	3.3933	3.3083	3.588*

Covariates - organizational tenure, gender and marital status
$p \leq .05$* $p \leq .01$** $p \leq .001$***

Table 4: Results of MANCOVA for work outcomes

Variables	Telecommuters (N=89)	Non-telecommuter (N=71)	F
Organizational Commitment	3.5478	3.5762	0.790
Job Satisfaction	3.8127	3.6479	1.114
Career Advancement Prospects	1.6966	1.5915	5.306**

Covariates - organizational tenure, gender and marital status

p ≤ .05*   p ≤ .01**   p ≤ .001***

reported experiencing more strain-based work-family conflict than non-telecommuters. Hypothesis 6, which proposed telecommuters would participate in fewer boundary-spanning activities than non-telecommuters, also was not supported. The telecommuters actually reported participating in more boundary-spanning activities than non-telecommuters.

Table 4 shows the results of the MANCOVA examining differences in work outcomes between telecommuters and non-telecommuters. None of the hypotheses concerning the impact of telecommuting on work outcomes were supported. Hypotheses 7 and 8 predicted differences between telecommuters and non-telecommuters on organizational commitment and job satisfaction. There is not a statistically significant difference in these work outcomes for telecommuters and non-telecommuters. Hypothesis 9 proposed that telecommuters would have fewer career advancement opportunities than non-telecommuters. This was not supported. In fact, after controlling for gender, marital status and organizational tenure, telecommuters reported higher career advancement prospects than non-telecommuters.

DISCUSSION

Many of the positive beliefs concerning telecommuting were found to be realities. The telecommuters reported significantly more autonomy then their non-telecommuting peers. By working outside of the traditional office environment, the telecommuters felt they had more discretion and control over the completion of their work. Relatedly, the belief that telecommuters would experience less time-based work-family conflict was found to be a reality. Because telecommuters have more control over when and where their work is completed, they are able to avoid or minimize time conflicts between work and family. For example, telecommuters can more readily restructure their day to attend a parent-teacher conference than employees who work in the traditional office environment.

The only negative belief that was confirmed by this research is that telecommuters receive less career support than their non-telecommuting peers. Providing less career support may be intentional on the part of the supervisor. Perhaps supervisors view telecommuters as being less committed to their careers and therefore less deserving of career support. Conversely, limited career support for telecommuters may be much less sinister and simply a function of limited interaction. By spending less time together in the office, telecommuters and their supervisors may simply have less time to talk about career development issues. Since career support can have a negative impact on advancement opportunities, this finding should be of particular concern to telecommuters and their supervisors.

Many have forecasted that telecommuting will result in negative experiences and outcomes, including more work-family conflict and less boundary spanning activities, organizational commitment and career advancement

Table 5: Supported and unsupported hypotheses

H1: Telecommuters will experience more autonomy than non-telecommuters	Reality
H2: Telecommuters will experience less time-based work-family conflict than non-telecommuters.	Reality
H3: Telecommuters will experience more strain-based work-family conflict than non-telecommuters.	Myth
H4: Telecommuters will experience more behavior-based work-family conflict than non-telecommuters.	Myth
H5: Telecommuters will have less career support than non-telecommuters.	Reality
H6: Telecommuters will have fewer boundary spanning activities than non-telecommuters.	Myth
H7: Telecommuters will have less organizational commitment than non-telecommuters.	Myth
H8: Telecommuters will have more job satisfaction than non-telecommuters.	Myth
H9: Telecommuters will have less career advancement prospects than their non-telecommuting peers.	Myth

prospects. All of these beliefs were found to be myths. Telecommuters did not report more strain and behavior-based work-family conflict than the non-telecommuters, suggesting that employees who work from home are able to sufficiently separate their work and personal lives. Telecommuters reported participating in more boundary-spanning activities than their non-telecommuting peers; therefore, telecommuters should not fear that working outside of the office a few days per week will hinder their visibility. There was not a significant difference in the level of organizational commitment reported by telecommuters and non-telecommuters. These results indicate that supervisors do not need to worry about the commitment level of their telecommuting employees. Finally, the concern that telecommuting will limit career advancement is unfounded. Despite receiving less career support, telecommuters did not report a negative impact on career advancement prospects. In fact, the telecommuters reported higher advancement opportunities than their non-telecommuting peers. This may occur because telecommuting employees are more productive and participate in more boundary-spanning activities and therefore have more visibility and advancement opportunities.

LIMITATIONS AND DIRECTIONS FOR FUTURE RESEARCH

Like all studies, these results are a function of the population that was sampled. Since the telecommuters and non-telecommuters were from one organization, the generalizability of these findings are limited. It is possible that these positive experiences and outcomes are limited to the organization that was examined and may or may not be similar for telecommuters from other organizations. Additional research, using a wider and larger sample, is necessary to confirm the generalizability of these findings.

One of the greatest myths that this research helps to dissuade is that telecommuters will forfeit their advancement opportunities. In fact, this research found telecommuters perceive their advancement opportunities to be significantly greater than non-telecommuters. This research used a self-assessment of career advancement prospects. Although other researchers have used self-assessment of career advancement prospects, the career advancement assessments of a supervisor may be a better predictor of actual advancement. Additional research should examine whether there are differences in the supervisors' assessment of career advancement for professional telecommuters and non-telecommuters.

CONCLUSIONS AND RECOMMENDATIONS

This study found the work experiences and outcomes for professional telecommuters to be very positive. The only negative finding in this study was that telecommuters reported experiencing less career support than non-telecommuters. Supervisors should be conscious of this potential negative work experience when managing telecommuting employees. Frequent communication and a conscious effort to support telecommuting employees should eliminate this concern.

This research was conducted with a sample from a telecommunications firm that was very proactive in developing flexible work arrangements. The design and implementation of the telecommuting program could have contributed to the positive experiences and outcomes identified in this research. Practitioners should consider three critical areas when designing a telecommuting program (Riley & McCloskey, 1997). First, management must support the program. As this research found, telecommuters may receive less career support. The lack of support for telecommuting from management would likely result in even less career support, isolation, dissatisfaction and limited career advancement prospects. Second, there should be a comprehensive, written telecommuting policy that covers legal concerns and human resource policies. Issues such as work hours, the use of personal equipment and reimbursement for work expenses should be clearly addressed. Finally, organizations should be cautious when selecting employees to telecommute. Everyone may not be successful in a flexible work environment. Typically this work option is best for high-performing employees who have been with the organization for at least one year. With careful design and implementation, telecommuting programs that allow organizations and employees to reap the many rewards of flexible work arrangements, while minimizing the potential disadvantages, can be a reality.

REFERENCES

Baroudi, J. K. (1985). The impact of role variables on IS personnel work attitudes and intentions. *MIS Quarterly, 9,* 341-357

Bailey, D. S. & Foley, J. (1990). Pacific Bell works long distance. *HRMagazine,* 35, 50-52.

Connelly, J. (1995). Let's hear it for the office. *Fortune,* 131, 22-23.

DiMartino, V., & Wirth, L. (1990). Telework: A new way of working and living. *International Labour Review,* 129, 529-554.

DuBrin, A. J., & Barnard, J. C. (1993). What telecommuters like and dislike about their jobs. *Business Forum,* 18, 13-17.

Fitzgerald, K. M. (1994). Telecommuting and the law. *Small Business Reports,* 19, 14-18.

Fuss, D. (1995). Telecommuting—The way of the future. *Telecommunications,* 15, 21.

Gaitley, N. J. (1996). *The influence of social support and locus of control on the well-being of men and women in the work-family domain.* Unpublished Doctoral Dissertation, Drexel University, Philadelphia, PA.

Gordon, G. E. (1998). Current telecommuting survey data shows strong growth. *Telecommuting Review,* 15(11), 1-3.

Greenhaus, J. H. & Beutell, N. J. (1985). Sources of conflict between work and family roles. *Academy of Management Review,* 10(1), 76-89.

Greenhaus, J. H, Parasuraman, S., & Wormley, W. (1990). Effects of race on organizational experiences, job performance evaluations and career outcomes. *Academy of Management Journal,* 33, 64-86.

Hackman, J. R. & Lawler, E. E. (1971). Employee reactions of job characteristics. *Journal of Applied Psychology,* 55(3), 259-286.

Hackman, J. R. & Oldham, G. R. (1975). Development of the job diagnostic survey. *Journal of Applied Psychology,* 60, 159-170.

Igbaria, M., & Greenhaus, J. H. (1992). Determinants of MIS employee's turnover intentions: A structural equation model. *Communications of the ACM,* 35, 34-49.

Keller, R. T., & Holland, W. E. (1975). Boundary-spanning roles in a research and development organization: An empirical investigation. *Academy of Management Journal,* 18, 388-393.

Lewis, H. (1997). Exploring the dark side of telecommuting. *Computerworld,* 31, 37.

McCloskey, D. W., & Igbaria, M. (1998). A review of the empirical research on telecommuting and directions for future research. In M. Igbaria & M. Tan (Eds.), *The Virtual Workplace.* Hershey, PA: Idea Group Publishing.

McShulskis, E. (1998). Telecommuting becomes a standard benefit. *HRMagazine,* 43, 28-29.

Miles, R. H., & Perrault, W. D. (1976). Organizational role conflicts: Its antecedents and consequences. *Organizational Behavior and Human Performance*, 17, 19-44.

Olson, M. H. (1989). Organizational barriers to professional telework. In E. Boris and C.R. Daniels (Eds.) *Historical and Contemporary Perspectives on Paid Labor at Home*. Urbana, IL: University of Illinois Press.

Olson, M. H., & Primps, S. B. (1984). Working at home with computers: Work and nonwork issues. *Journal of Social Issues*, 40(3), 97-112.

Parasuraman, S., Greenhaus, J., & Granrose, G.S. (1992). Role stressors, social support and well being among two career couples. *Journal of Organizational Behavior*. 13(4), 339-356.

Porter, L. W., Crampon, W. J. & Smith, F. J. (1976). Organizational commitment and managerial turnover: A longitudinal study. *Organizational Behavior and Human Performance*, 15, 87-98.

Riley, F., & McCloskey, D.W. (1997) Telecommuting as a response to helping people balance work and family. In J. Greenhaus and S. Parasuraman (Eds.) *Integrating Work and Family: Challenges and Choices for a Changing World*. Wellsport, Conn: Quorum Books.

Stanko, B. B. & Matchette, R. J. (1994, October/November). Telecommuting: The future is now. *B&E Review*, 8-11.

Wright, P. C. (1993). Telecommuting and employee effectiveness: Career and managerial issues. *International Journal of Career Management*, 5, 4-9.

Young, W., & Morris, J. (1981). Evaluation by individuals of their travel time to work. *Transportation Research Record*, 794, 51-59.

About the Authors

Nancy Johnson is the associate dean of the School of Business at Capella University. She had been an assistant professor of MIS with Metropolitan State University, Minneapolis, MN from 1991-2000. She was in private industry for twenty years prior to that. Author of four book chapters and numerous articles, and conference presenter, she was also a guest editor for a special edition on telecommuting of the *Journal of End User Computing*. She has also been a book and Website reviewer for *Choice Journal* for five years. She was a Fulbright Scholar in 1992 in Malaysia. She holds a BS and an MBA from University of Minnesota and a PhD from Walden University. Her research interests include distance education, human factors in successful change management, justification methodologies of IT investment in the public sector, and international use of IT. (njohnson@capella.edu)

* * *

Frédéric Adam is a lecturer in the Department of Accounting, Finance and Information Systems at University College Cork in Ireland. He is also a senior researcher with the Executive Systems Research Centre (ESRC). He holds a PhD from the National University of Ireland and Université Paris VI jointly. His research has been published in a number of international journals including the *Journal of Strategic Information Systems*, *Decision Support Systems* and *Systèmes d'Information et Management*. He is the co-author of the *Manager's Guide to Current Issues in Information Systems* and the *Practical Guide to Postgraduate Research in the Business Area* (both published by Blackhall Publishing, Dublin, Ireland).

Dr. Yehuda Baruch is a senior lecturer at the School of Management, UEA, Norwich, UK, and formerly a visiting research fellow at London Business School. He earned a DSc in management and behavioral sciences from the Technion, Israel. His research interests are human resource management, mainly career management systems, teleworking, and computer-mediated communication. He has published in these fields in a number of journals, including *Human Relations, Organizational Dynamics, Human Resource Management* and *Organization Studies.*

Martin Boisvert is the vice-president of S.I.X. Inc. in Montreal.

Joseph R. Bumblis is currently a network management architect with U S WEST in Minneapolis, Minnesota. Previous positions include manager of systems software development at Seagate Technology, systems engineer/ systems architect for Control Data Corporation, computer communications design specialist at McDonald Douglas Space Systems Company, a researcher and senior member of technical staff at MCC, and a development program manager at Tandem Computers.

Joe received a master of science degree from the University of Minnesota in management of technology, and a bachelor of science degree from Ohio University in electrical engineering. Joe is active in the IEEE and as an adjunct instructor of information technology and management for Concordia University and Metropolitan University in Minneapolis, MN.

Gregory Crossan is an associated researcher of the Executive Systems Research Centre (ESRC) at University College Cork in Ireland. After successfully completing his master of business studies in management information and managerial accounting systems (MIMAS) in the Department of Accounting, Finance and Information Systems, he has taken up employment in an Irish software company and develops business solutions for the World Wide Web. His research interests include the Internet and the World Wide Web and the use of IT for a better integration of the disabled in the work environment.

Kitty de Bruin is the managing director of TelewerkForum (www.telewerkforum.nl). She can be reached at e-mail: kitty.de.bruin@telewerkforum.nl.

Susan J. Harrington is a professor in information systems at the J. Whitney Bunting School of Business at Georgia College & State University, Milledgeville, GA. Her interests include ethical decision making in organizations, IT adoption and diffusion, and corporate culture. She has published in *Journal of Business Ethics, MIS Quarterly, DATA BASE for Advances in Information Systems, IEEE Transactions on Professional Communication* and *Academy of Management Executive*, as well as other management journals.

Janet A. Henquinet received her PhD from the University of Minnesota and is an associate professor of management at Metropolitan State University in Minneapolis, MN. Prior to joining Metropolitan State she held a variety of human resource management positions in private and public sector organizations. She currently teaches graduate and undergraduate courses in management, organizational behavior and human resource management with a special interest in international human resource management.

Donna Weaver McCloskey is a graduate of the University of Delaware (BS), Widener University (MBA) and Drexel University (PhD). She is currently an assistant professor of MIS in the School of Business Administration at Widener University. Her research interests include technology acceptance, data warehousing and the impact of technology on changing work roles. Dr. McCloskey's research has been published in a number of books and journals and has been presented at international and national conferences.

Diana Page, PhD, is an associate professor with the Department of Management & MIS at the University of West Florida. Her primary teaching areas include organizational behavior and business communication. Her work experience includes supervisory and management positions in the retail trade, human resources and the public school system. Dr. Page frequently consults with organizations such as Bell South, Instrument Control Service (a division of General Electric), Health & Rehabilitative Services, local hospitals and utility companies. Her current research interests include virtual teams, careers, gender issues and management pedagogy. She is published in journals such as the *Journal of Social Behavior and Personality, Education,* the *Journal of Business Ethics*, and the *Organizational Development Journal*.

Alain Pinsonneault holds a PhD from the University of California at Irvine and is the Imasco Associate Professor of Information Systems in the Faculty of Management at McGill University. Prior to joining McGill University in June 1999, he was an associate professor at the École des Hautes Études Commerciales (HEC-Montreal) and the director of the PhD program.

Alain Pinsonneault won numerous awards for his research and his teaching, among which is the Doctoral Award of the International Center for Information Technology and MCI Communications in the USA in 1990. His current research interests include the organizational and individual impacts of information technology, electronic commerce, group support systems, and IT department management. He has published papers in *Management Science, Management Information Systems Quarterly, Information Systems Research, the Journal of Management Information Systems, Decision Support Systems*, and the *European Journal of Operational Research*.

Richard Platt, PhD, is an assistant professor of management information systems at the University of West Florida in Pensacola, Florida. His primary teaching areas include strategic planning and management of information systems, electronic commerce, and information systems design and development. He is published in the areas of electronic commerce, competitive uses for information systems and technology, managing technology enhanced teams, and critical success factors for IS projects. He consults in the areas of database analysis and design, and the planning, design, and development of all levels of information systems in both public and private IS organizations.

Jay Rodstein is a principal environmental, health and safety engineer for Honeywell. He has worked at the Honeywell Technology Center for seven years. He holds a bachelor of arts degree in chemistry from Western Maryland College and a master's in business administration from the University of Pittsburgh. Jay has worked in the environmental and safety fields for 19 years and is a certified hazardous materials manager (CHMM).

Cynthia P. Ruppel is an assistant professor in information systems in the College of Business at The University of Toledo, Toledo, OH. Her interests include telecommuting and other forms of virtual work, IT innovation, adoption and diffusion, as well as organizational culture and change management. She has published in *DATA BASE for Advances in Information Systems, Journal of End User Computing, IEEE Transactions on Professional Communication* and *Information Resource Management Journal*.

Ian Smith is the Clifford Chance Professor of Employment Law at the University of East Anglia, UK. He is a barrister with Chambers in the Middle Temple in London, specializing in employment law and health and safety issues. He is also an industrial arbitrator for ACAS. Among his numerous books is *Smith and Wood Industrial Law*, the leading textbook in this area; and he is an editor of *Harvey on Industrial Relations and Employment Law*, the standard practitioner reference work.

Kirk St. Amant has a BA in anthropology and in government from Bowdoin College and an MA in technical and scientific communication from James Madison University. He is currently enrolled in the University of Minnesota's PhD program in rhetoric and technical and scientific communication where his research interests include international and intercultural communication and communication in the on-line environment. In addition to being a student, he does independent consulting work on international communication and localization issues. He can be reached at stam0032@tc.umn.edu.

Sandy Staples, PhD, is an assistant professor of management information systems in the School of Business at Queen's University, Kingston, Canada. His research interests include the enabling role of information systems for virtual work and knowledge management, business process reengineering, and assessing the effectiveness of information systems. Dr. Staples has published articles in: *Organization Science*, *Journal of Business and Psychology*, *Communications of the Association of Information Systems*, *Journal of Computer-Mediated Communications*, *Australian Journal of Information Systems*, *Business Quarterly*, *International Conference on Information Systems*, *Journal of End-User Computing* (forthcoming), *Journal of Strategic Information Systems* (forthcoming), and *OMEGA*.

Kate Watters is an occupational health nurse for Honeywell. She received her bachelor of science degree in nursing from St. Olaf College in Northfield, MN. She is a registered nurse and has been in the occupational health field for 16 years. She has worked at Honeywell for 10 years. She is currently managing the occupational health department at the Honeywell Technology Center in Minneapolis, MN.

Index